Excel Manual
for
Moore and McCabe's

Introduction to the Practice of Statistics
Fifth Edition

Linda Getch Dawson
University of Washington Tacoma

W. H. Freeman and Company
New York

Printed in the United States of America

ISBN: 0-7167-6360-5
EAN: 9780716763604

First printing

W. H. Freeman and Company
41 Madison Avenue
New York, NY 10010
Houndmills, Basingstoke
RG21 6XS England

TABLE OF CONTENTS

Preface

This manual is a supplement to the textbook, *The Introduce of the Practice of Statistics,* by David S. Moore and George P. McCabe, also referred to as IPS. It aids the reader in performing statistical analysis using Excel. In addition, I wrote two macros, in conjunction with writing this book, to enhance the statistical and graphical analysis capabilities of Excel and to carry out specific analysis required in the IPS textbook. The output for these procedures is easy to read and interpret. The programmed macros are available online at the IPS Web site (http://bcs.whfreeman.com/ips5e/) with installation instructions in the *Introduction* of this book.

I have taught different levels of college statistics for almost 15 years, using almost exclusively Moore authored textbooks. I have also used a variety of statistical analysis programs—Excel, Minitab, SPSS and others. Excel has worked the best as an accessible computer tool capable of most basic to intermediate analysis. In my experience, students appreciate that the statistical methods learned using Excel as the analysis tool are easily transferable to work and research outside of the classroom.

Key features of this book include:

- Introductory chapter for Excel covering basic program operation and procedures.

- Step-by-step solutions with illustrative figures for examples used in every chapter of the IPS textbook.

- Summary content from the IPS text is provided as introductory material and explanations for concepts in each chapter. Many of these comments and explanations are closely aligned with the writing in the text to serve as a background for the examples worked in Excel.

I would like to personally thank the team members at W. H. Freeman and Company who were involved in this project including: Craig Bleyer, Amy Schaffer, and Victoria Anderson.

Finally, I would like to thank my husband Allan for giving up countless weekends and provided a nurturing environment throughout this process. Thank you also to my sisters, Judith Getch Brodman and Patricia Getch for providing endless moral support for all of my pursuits. And, finally, to my parents, Charles and Katherine Getch, for encouraging me to achieve excellence in education.

Linda Getch Dawson
University of Washington Tacoma
Tacoma, Washington
January 15, 2005

Introduction

This manual, a supplement to the text *Introduction to the Practice of Statistics* by David S. Moore and George P. McCabe, also referred to as IPS, will aid the reader in performing the statistical analysis needed to solve the examples detailed in IPS. **Excel** and a couple of programmed **macros** available on the IPS companion Web site (http://bcs.whfreeman.com/ips5e/) will be used to perform the analysis. Data sets are available on the Web site as well as on the CD-ROM accompanying the IPS textbook.

I.1   Using Excel

Microsoft Excel is a widely used spreadsheet application that includes capabilities such as spreadsheet analysis, graphic display, and database applications. An electronic spreadsheet organizes data much like an accounting worksheet or table of data. Excel can also perform statistical analysis using its built-in functions and other tools.

Macros were developed by the author of this manual to assist in creating boxplots and normal quantile plots, a capability that does not exist in Excel. The macros are available on the textbook companion Web site (http://bcs.whfreeman.com/ips5e/) and can be downloaded for use on your own computer.

Versions of Excel

The examples in this book were written using Excel 2000 running on Microsoft Windows 2000 Professional. The code successfully operates under all subsequent versions of Excel and Windows XP. Compatibility should also exist with the Macintosh applications but has not been tested.

Prior Knowledge

It is not necessary to have any prior knowledge of Excel to use this book. However, it is helpful to practice some of the program's basic functions outlined in the next section before using the program for statistical analysis.

I.2   Worksheet Basics

When you open Excel, a new file opens with a series of blank worksheets organized into a workbook. Initially, three worksheets are available to you, but up to a total of 255 sheets can be created. You can also insert, move, or delete worksheets as needed. The figure below is a screenshot of a Microsoft Excel 2000 worksheet.

The outer window is the Excel program window. The inside window is the workbook. In this case, both windows have been maximized. The worksheet tabs are shown at the bottom of the workbook. Each sheet is selected by clicking on its tab. The worksheets can be renamed and reorganized within a given workbook or between other open workbooks.

Worksheet data is displayed in cells, which can be identified by the row and column that intersect at that point. For example, the first cell is named A1 and is located at the intersection of column A and row 1. There are 65,536 rows and 256 columns in a single worksheet. The rows are numbered on the left and the columns are labeled at the top. The columns start with A and increase alphabetically to Z, starting over with AA, AB, AC, and continuing up through IV. Cell data is also referenced by the specific sheet in which the data is input for instances where data or formulas are shared among worksheets.

Data is entered into a cell by selecting a cell address either by left-clicking on it with the mouse or by moving to that location using the arrow keys. Information can be typed directly into a particular cell or in the Formula Bar above the actual worksheet. Three types of information can be entered into a cell: labels, values, and formulas. The length of a cell's content can be quite long; however, the information may not be displayed in the cell if it is too long or if the column is not widened. The maximum column width is 255 characters.

I.3   Components of the Workbook

Screen Information

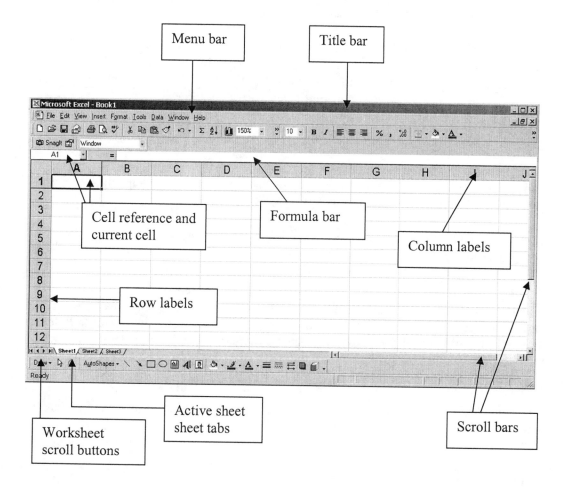

Excel Menus

Excel menus drop down after you click on the menu heading, revealing submenus where appropriate. If a drop down menu reveals an arrow, place the cursor on that line to view the remaining items in that menu choice.

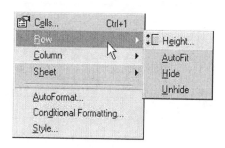

The menu bar pull-down options are fully defined in the table below:

Menu	Pull-down options
File	Open, close, save, page setup, print, exit
Edit	Copy, cut, paste, delete, move
View	Controls what elements are viewed on screen
Insert	Insert rows, columns, sheets, text, etc.
Format	Format cells, rows, columns, sheets
Tools	Access tools in Excel such as macros, data analysis toolpak (used extensively in this manual)
Data	Database functions such as sorting, filtering
Window	Organize and display options for worksheets
Help	Built-in help features

In addition to the pull-down menu options, there are keyboard equivalent controls and right-click menu options. Often, there is more than one way to accomplish a task. Use whatever method is easiest for your own working style.

The Formula Bar

The **Formula Bar** is located above the worksheet. Cell contents can be input using the **Formula Bar**. Its unique features include access to the **Formula Palette** when you type or click the equal sign and then on the drop-down arrow on the left. The **Formula Palette** is a built-in tool in Excel that assists with function syntax.

H23	▾	✗ ✓ =	

The **cancel X** and the **accept check mark** appear only when information is being input or edited into a cell.

Sheet Tabs

Each worksheet has a tab located at the bottom of the worksheet. Clicking on an individual tab activates that sheet. More than one sheet can be activated by selecting the first sheet, by holding the **Control** key down and then selecting another sheet(s). You can click and drag a sheet tab to move the sheet to another location in the worksheet order. By right-clicking on the sheet tab and selecting **Rename**, the worksheet can be named to reflect its contents. A new worksheet can be added by selecting **Insert** ⇨ **Worksheet** from the menu.

37					
38					
39					
40					
41					

◄◄ ◄ ► ►►\ **Sheet1** ⟨ Sheet2 ⟨ Sheet3 ⟩

I.4   Entering Data

When a new workbook opens, cell A1 automatically becomes the active cell, with a dark outline surrounding it. Click with the mouse or move with the arrow keys to activate a different cell. When data is input, you can press the **Enter** key on the keyboard or click on the **Accept** box to finalize the cell entry.

You can enter several types of data into a worksheet cell: Text, numbers, and formulas are the most common. Before you begin, you must select the cell in which you want to enter the information.

Labels

A label is text that can contain letters, numbers, or special characters. The text is, by default, left aligned in the cell. This can be changed through the toolbar or menu if desired. The number of characters displayed and printed in a cell depends on the width of the column. If the cells to the right are empty and remain empty, you can enter a long label, up to a maximum of 255 characters.

Numbers

The rules for entering numbers are as follows:
- No spaces
- First character must be 0–9, +, − , $
- Can include a comma or %
- Negative numbers are entered by typing a dash first

The numbers, by default, are right aligned. Once the number is entered, you can format it (font, alignment, number of decimals) to change its appearance.

Formulas

Formulas are mathematical expressions that can use values or formulas in other cells to calculate new values. Formulas can include numbers, cell locations, ranges, functions, and labels. Once the formula is entered and accepted, the result will be displayed in the cell and the equation will be displayed in the **Formula Bar.**

To create a formula, the computer must recognize that you are entering a formula. An "=" sign is placed before the formula so that the computer can identify that there is a formula in the cell. Cell locations are used in formulas to give your worksheet more flexibility.

The four major operations are:
- ♦ Addition   +
- ♦ Subtraction   –
- ♦ Multiplication   *
- ♦ Division   /

Functions

Functions are special formulas that perform more complex operations. They are pre-programmed equations in Excel designed to save time and increase efficiency. Functions can be used for arithmetic, statistical, scientific, or investment calculations. The most common functions are the SUM, AVG, MAX, and MIN, but there are dozens of functions available for your use. You can input a function by typing it directly into a cell or by accessing the **Formula Palette** or **Function Wizard** for assistance.

An example of a function is the calculation of a mean or average of a data set. The function is input into the cell where you want the answer. All functions start with "=" as other formulas. The AVERAGE function is of the form **"=AVERAGE (A2:A25)"** where the range in the parentheses represents the data range of which you want to find the average, selected manually by dragging with the mouse. The function name is not case sensitive.

I.5   Modifying Data

Editing

There are several ways to modify or edit the information in a cell. For data that has not yet been entered after typing, you can use the **Backspace** or **Delete** key to modify the contents. You can also press the **ESC** key to cancel the entire operation or click on the **X** box in the **Formula Bar**.

Once data has been entered into a cell, you can double-click on the cell and move the cursor I-beam to allow editing. You can also activate the cell and edit the data in the **Formula Bar**.

Selecting Cell Ranges

You can click and drag with the mouse (hold the left key on the mouse down while moving the cursor) to select a cell or a group of adjacent cells, defined as a range. To select multiple ranges, select the first range, hold the **Control** key down, and select any other ranges of cells. The worksheet shown on the next page illustrates four types of ranges: an individual cell, a row of cells, a block of multiple columns and rows, and a column group. A range must be a rectangular shape or a group of adjacent cells. A range designation is defined as the top, upper, or leftmost cell location, a colon, and the last cell on the bottom or the right of the range. Ranges are selected to change the formatting of a worksheet or to calculate formulas.

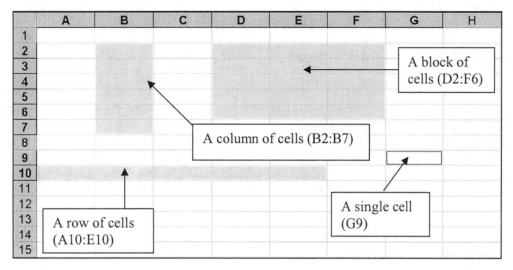

Ranges can be named for ease of use in complex spreadsheets with multiple references.

Deleting or Clearing Data

To delete cell contents, select the cells that you want to delete and then press the **Delete** key. There is the option to remove the cell's formulas (contents), formats, comments, or all. Comments are notes attached to cells and are used to document a worksheet. They appear when you point to the cell.

To clear cell contents, select the cells that you want to clear and then select **Edit** ⇨ **Clear** from the menu. The options are defined as follows:

All	Erases everything in the selected area
Formats	Erases only the formats, leaving the contents and comments
Contents	Erases only the contents, leaving the formats and comments
Comments	Erases only the comments

Inserting Rows and Columns

To insert a row, click on the row heading (the number) where you want to add a row and select **Insert** ⇨ **Rows.** A new row is inserted and any information previously in that row is moved down. To insert more than one row, click and drag on more than one row heading (be sure to click and drag in the center of the row heading button, or you could resize the row) and select **Insert** ⇨ **Rows.**

To insert a column, click on the column heading (the letter) and then select **Insert** ⇨ **Column.** To insert more than one column, click and drag on the column headings (be sure to click and drag in the center of the column heading button or you could resize the column) and select **Insert** ⇨ **Column.**

Deleting Rows and Columns

To delete one or more rows, click on one or more row headings and press the **Delete** key on the keyboard, or select **Edit** ⇨ **Delete** from the menu. The entire row is deleted, including all the information in that row.

To delete one or more columns, click on one or more of the column headings and press the **Delete** key on the keyboard or select **Edit** ⇨ **Delete** from the menu. The entire column is deleted, including all the information in that column.

Moving Information

You can move any information in your worksheet from one section to another either on the same sheet or a different sheet, or even to a different workbook or application.

Cut and Paste

If you want to **Cut** and **Paste** information, select the range of information that you want to move and select **Edit** ⇨ **Cut** from the menu or click the **Scissors** or **Cut** button ✂ in the toolbar. Select the cell that represents the upper left cell in the new location and then select **Edit** ⇨ **Paste** from the menu or click the **Paste** button 📋 in the toolbar. You should be aware that certain types of formulas or functions might not behave the way you expect them to when they are cut and pasted into a new cell.

Copy and Paste

If you want to copy selected information instead of moving it, select the range of information that you want to copy and select **Edit** ⇨ **Copy** from the menu, or click the **Copy** button 📋 in the toolbar. Select the cell that represents the upper left cell of the new location area and select **Edit** ⇨ **Paste** from the menu, or click the **Paste** button 📋 in the toolbar. You should be aware that certain types of formulas or functions might not behave the way you expect them to when they are copied and pasted into a new cell.

Drag and Drop

If you want to drag and drop individual cells or a range of cells from one location to another, select the range of information that you want to move and then place the mouse pointer on the edge of the cell or range until you see an arrow-shaped pointer. At that spot, you can click and drag the range of data to another location.

Filling

If you want to fill a range of cells in with a specific number or a series of numbers, type and enter the data to be repeated. Select the cell(s) to be copied and then move the cursor to the lower right of the cell until the mouse pointer turns into a cross hair. Drag in whatever direction you want to copy the information. If an extended series is desired, type the first two numbers or time increments in the series, then select those two cells and drag down the lower right cross hair until the series is completed.

	A	B	C
1			
2		10	
3		20	
4			
5			30

Formatting a Worksheet

You can change the appearance of individual cells or a range of cells. For example, you can enhance the text by changing the font size, change the look of a number to display a certain number of decimal places or a "%" or "$", or add color to the worksheet. Select the cell(s) to be formatted and then select **Format** ⇨ **Cells** from the menu and choose the desired options. Additional options can be selected from the formatting buttons in the toolbar.

Adjusting Column Width and Row Height

The width of one or more columns can be adjusted by clicking and dragging on the line between the column heading that you are adjusting and the next column. You will see an adjust symbol on that line. Drag to the right to widen. The column is adjusted when you let go of the mouse button. Alternatively, you can double-click on that same line for a "best fit" option for the entire column. The column can also be adjusted by clicking on one or more of the column headings and selecting **Format** ⇨ **Column** ⇨ **Width** from the menu.

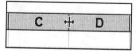

The height of one of more rows can be adjusted by clicking and dragging on the line between the row heading that you are adjusting and the next row. You will see an adjust symbol for the mouse on that line. Drag to increase the height. The row is adjusted when you let go of the mouse button. The row can also be adjusted by clicking on one or more of the row headings and selecting **Format** ⇨ **Row** ⇨ **Height** from the menu.

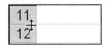

Number Formatting

The number format determines how the numbers appear on a worksheet, both on the screen and when printed. The numbers are formatted with a **General** type format by default. You can choose from a variety of formats by selecting the cells to change, **Format** ⇨ **Cells** from the menu, and then the **Number** tab. The format command does not change or round off the number; it only changes the appearance of the number.

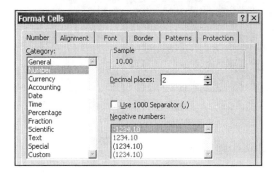

Changing Fonts

You can change the look of your worksheet by changing the font type and font size. Row height and column widths are automatically adjusted to accommodate font changes unless you have previously set the row to a specific height.

Select the range of cells that you want to change, **Format** ⇨ **Cells** from the menu, and then the **Font** tab. You can also use the formatting buttons in the toolbar.

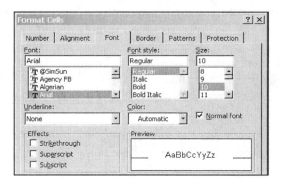

I.6 Printing

Print Preview

To view your worksheet before you print, click the **Print Preview** button 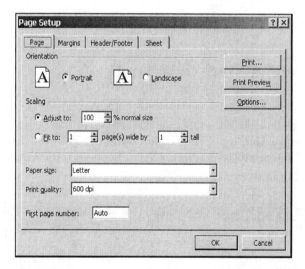 in the toolbar. The preview displays the current page and shows you the general layout. You can zoom in by clicking with the mouse pointer. There are also options to control margins, page setup, and printing.

Print

You can print one copy of your current worksheet by clicking on the printer button in the toolbar. Printing options include printing the entire workbook or only selected sheets, or printing selected or highlighted section(s) in your workbook. You can also print only certain pages and change the number of copies to be printed. Select **File** ⇨ **Print** for more options.

Page Setup Options

The **File** ⇨ **Page Setup** menu option is used to set or change how the finished page looks when viewing or printing.

Options can be selected for adjusting the page, margins, headers/footers, and the worksheet orientation.

I.7 Help

This introductory chapter gives you a brief overview of Excel and its capabilities. Specific methods, as they relate to graphic and statistical analysis, are detailed in the following chapters. However, if you are a novice to Excel, it is recommended that you gain expertise through practice with a tutorial-related basic Excel text or a training workshop. There are numerous texts that can provide useful information, but nothing replaces practice.

An extensive help system is provided with Excel. You can access help by clicking on the **?** button in the menu bar. You can select your own animated helper and ask specific questions. There is also a search capability with the **Help** menu.

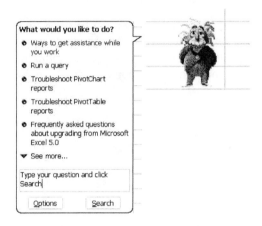

I.8 Using Excel's Statistical Tools

Excel contains a set of built-in statistical analysis tools that can be accessed by selecting **Tools** ⇨ **Data Analysis** from the menu. If that option is not available, select **Tools** ⇨ **Add-Ins** and then select **Analysis Toolpak.** Click **OK.** The **Data Analysis** option should now be available in the **Tools** menu.

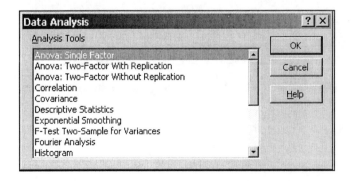

Many of the built-in Excel statistical analysis tools are detailed where appropriate in the exercises in the following chapters. The Excel statistical tools are useful for basic statistical analysis. However, the limitations include a lack of basic help when using statistical methods, non-intuitive output, and some missing procedures and tests (as well as errors) in more advanced analysis. Overall, however, it is a very powerful and useful program for business- and spreadsheet-related applications.

Exercises taken from the IPS text are solved using both the Excel analysis tools and the boxplot and normal quantile plot macros that are available to be downloaded from the IPS companion Web site (http://bcs.whfreeman.com/ips5e/).

The **macro** solutions are identified by the icon .

I.9   Using the Boxplot and Normal Quantile Macros

Macros have been programmed by the author of this manual and are available to use on the IPS companion Web site (http://bcs.whfreeman.com/ips5e/). Select **Excel Macros.** The **Creating a Boxplot** and **Creating a Normal Quantile Plot** macros can be downloaded by right-clicking on each one separately and selecting **Save Target As** from the menu. Select a folder or the desktop to which the file will be downloaded. Click **Save** and the macro will be downloaded to the chosen location. The macro can then be accessed at any time by opening the worksheet containing the macros.

Macros can only be downloaded to Excel if the security level is at medium or lower. This security level will allow the user the option to choose whether or not to run a macro. Select **Tools** ⇨ **Macro** ⇨ **Security** from the menu and then select **Medium.** Click **OK.**

The macro worksheet will be blank except for a button that activates the macro when clicked. Specific directions for use are contained in the exercises appropriate for that macro. Data sets can be copied and pasted onto the macro worksheet for analysis.

	A	B	C	D	E
1					
2					
3					
4					
5					
6			Create Boxplot		
7					
8					

Chapter 1

Looking at Data—Distributions

A first step in the study of statistics is the examination of data. A data set contains information about individuals or units. The information measured is in the form of a **variable,** defined as any characteristic of an individual. **Variables** are categorized as **categorical,** placing an individual in a group or category, or **quantitative,** a measurement of numerical value. Investigating the distribution of a variable is a logical first step in the analysis of data.

1.1 Displaying Distributions with Graphs

The process of **exploratory data analysis** usually starts with graphic displays accompanied by numerical analysis. It is helpful to examine each variable individually and then study the relationship between variables. The distribution of a categorical variable identifies categories and uses counts or percentages of individuals placed in a category for comparison.

Categorical Variables: Bar Graphs and Pie Charts

Categorical variables can be displayed by using either a **bar graph** or a **pie chart.** Sizes of groups can be easily compared using a **bar graph.** A specific type of **bar graph** where categories or groups are ordered from most frequent to least frequent are called **Pareto charts. A pie chart** shows each group as a part of the whole.

IPS Figure 1.1

How well educated are 30-something young adults? Here is the distribution of the highest level of education for people aged 25 to 34 years.

Education	Count	Percent
Less than high school	4.6	11.8
High school graduate	11.6	30.6
Some college	7.4	19.5

Associate degree	3.3	8.8
Bachelor's degree	8.6	22.7
Advanced degree	2.5	6.6

Excel is capable of creating several types of charts (graphs) through the **ChartWizard.** The **ChartWizard** can be activated from the 📊 button in the standard toolbar or by selecting **Insert** ⇨ **Chart** from the menu.

1. Insert the following data into an Excel worksheet with no empty rows or columns.

	A	B	C
1	**Education**	**Count (millions)**	**Percent**
2	Less than high school	4.6	11.8
3	High school graduate	11.6	30.6
4	Some college	7.4	19.5
5	Associate degree	3.3	8.8
6	Bachelor's degree	8.6	22.7
7	Advanced degree	2.5	6.6

2. Adjust the columns to fit the width of the labels and headings. The method is to put your mouse pointer on the line between the column labels and then click and drag until you read the desired width. Double-clicking on the line will widen the column to the widest label in that column.

	A	↔
1	Education	
2	Less than high school	
3	High school graduate	
4	Some college	
5	Associate degree	
6	Bachelor's degree	
7	Advanced degree	

3. The data that we want to graph is the **Education** column and the **Percent** column, without the headings. These columns are separated by the **Count** column, which we don't want to use. To select two non-adjacent columns of data, select the first column, hold the **Control** (Ctrl) key down and select the second column.

	A	B	C
1	**Education**	**Count (millions)**	**Percent**
2	Less than high school	4.6	11.8
3	High school graduate	11.6	30.6
4	Some college	7.4	19.5
5	Associate degree	3.3	8.8
6	Bachelor's degree	8.6	22.7
7	Advanced degree	2.5	6.6
8			

4. Click the **ChartWizard** 📊 in the toolbar. The **Chart type** options are displayed in **Step 1**. The **Column** chart is the traditional bar graph with the bars displayed vertically. You can also select different **Chart sub-types** on the right side, including

a 3-D column chart. For this example, we will accept the default two-dimensional sub-type by clicking **Next.**

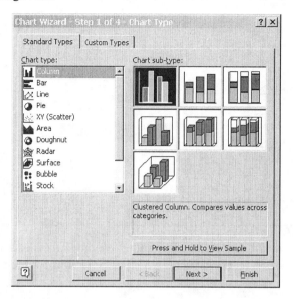

5. **Step 2** of the **ChartWizard** displays a preview of the chart, identifies the **Data Range** selected, and shows whether the **Series** is in rows or columns. The **Data Range** and **Series** location can be changed if the preview does not reflect the desired chart. The labels may appear slanted in the preview but can be changed after the chart is finished. Click **Next.**

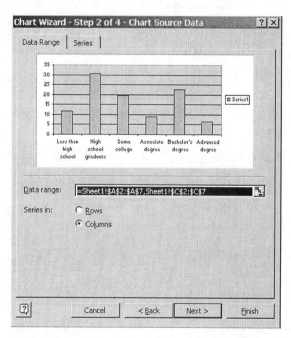

6. **Step 3** of the **ChartWizard** allows you to input **Chart Options,** including the **Chart Title** and the **Axis Labels.** For the **Chart Title,** input "Education Levels for Age 25

to 34 Years". Input the **Category (X) Axis label** as "Highest Education Level" and the **Value (Y) Axis label** as "Percent of People Aged 25 to 34".

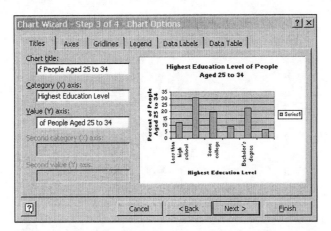

7. Several other **Chart Options** can be changed on the **Tab** sheets of the dialog box. Click the **Gridlines** tab and de-select the **Major Gridlines** for the *y* axis.

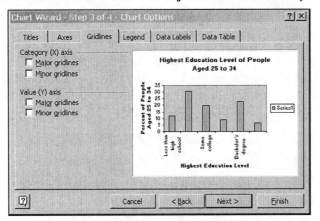

8. Click the **Legend** tab to de-select **Show Legend.** It is not necessary to display the legend if only one series of data is graphed. Click **Next.**

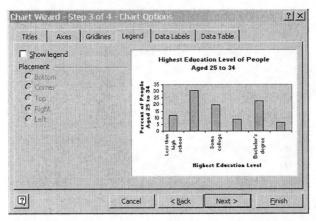

9. **Step 4** of the **ChartWizard** has two options for **chart location**–a **New Sheet** (chart is placed on its own worksheet) or an **object in** the current worksheet or another

worksheet. A chart placed on its own worksheet is more professional looking in documents and presentations. A chart placed as an object in a worksheet is appropriate for practice, homework problems, and instances where it is desirable to have the data and the chart next to each other. A chart placed as an object in a worksheet generally needs to be resized and fonts reformatted to have the best appearance. For this example, select **As new sheet** and click **Finish.** The resulting chart is shown on its own worksheet entitled **Chart 1.**

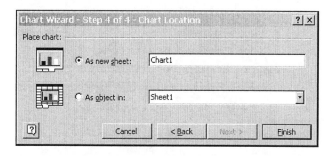

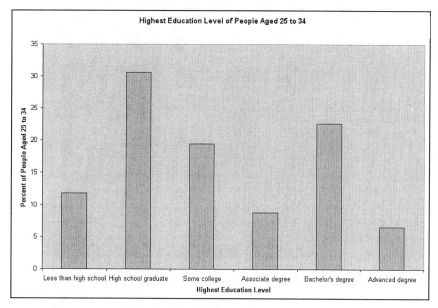

10. The resulting chart can be changed by going back into the **ChartWizard** by first clicking once on the chart to select it and then clicking the **ChartWizard** in the toolbar. Click **Next** to go to any other steps to make changes and then click **Finish.** Parts of a chart can also be changed once the chart is finished. See Excel **Help** for changes to color, gridlines, scales, etc.

11. A **Pareto chart** of the same data would require that the data be sorted from highest to lowest. Using the **Data** ⇨ **Sort** menu option, you can select sorting by **Percent.** Then follow the same instructions for creating the bar chart. If you sort your original data set, any charts that you have already created using this data set will be altered. If this is not desired, copy and paste the data set to another location and create another chart.

12. A **pie chart** of the Education data is created by selecting the Education data and the Count data. These columns are adjacent, so you can click and drag the data as one range and then click the **ChartWizard** 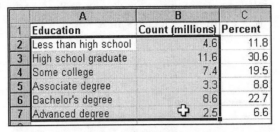 in the toolbar. Select **Pie** as the **Chart type** and the default **Chart sub-type** by clicking **Next**.

	A	B	C
1	**Education**	**Count (millions)**	**Percent**
2	Less than high school	4.6	11.8
3	High school graduate	11.6	30.6
4	Some college	7.4	19.5
5	Associate degree	3.3	8.8
6	Bachelor's degree	8.6	22.7
7	Advanced degree	2.5	6.6

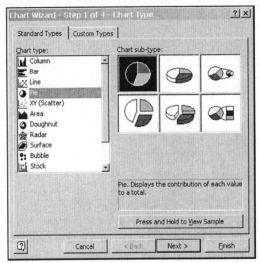

13. The chart preview in **Step 2** shows the pie chart with labels shown in a legend. This can be changed in the next step, so click **Next**.

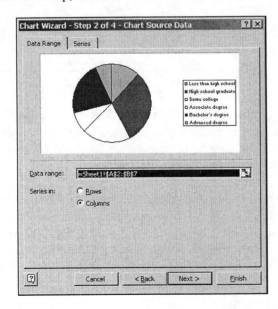

14. Input "Highest Education Level for People Aged 25 to 34" as the **Chart Title** and then click the **Legends** tab and de-select the legend. Click the **Data Labels** tab and select **Show label and percent.** Other versions of Excel may have a slightly different method for this step. Notice the other options available, including the already selected **Show leader lines.** Click **Finish.**

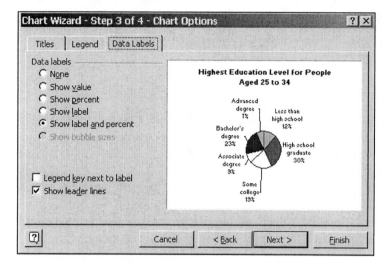

15. The chart is now located as an object on the data worksheet, which is the default. The chart needs to be resized by selecting it (click on it once) and then clicking and dragging the corners until all the labels can be seen clearly. Notice that the **Show label and percent** option automatically calculated the percentages, which should be the same as the data shown in the **Percent** column.

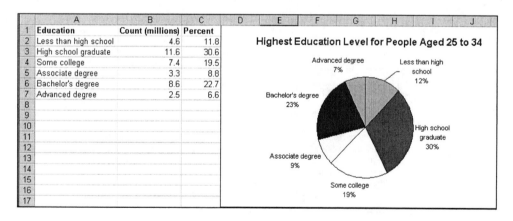

16. Save the file as **Education Graph Example.** It is assumed that you will save the remaining examples in this manual using appropriate names.

Quantitative Variables: Histograms

The distribution of a single quantitative variable can be illustrated by a **histogram,** a customized bar graph with values grouped together by intervals. A histogram has bin intervals charted on the *x* axis that are of equal width. Data values are placed into appropriate bins or groupings, much like sorting coins. The count of how many data points are in each bin is graphed on the *y* axis. The resulting histogram is uniquely different from any other graph and displays data distribution in a clear and descriptive way.

IPS Example 1.4 Data Analysis in Action: Don't Hang Up on Me

Many businesses operate call centers to serve customers who want to place an order or make an inquiry. Customers want their requests handled fully. Businesses want to treat customers well, but also want to avoid wasting time on the phone. So it is usual to monitor the length of calls and encourage representatives to keep calls short.

Table 1.1 displays the lengths of the first 80 calls from a data set containing 31,492 calls made to the customer service center of a small bank in one month.

Before creating a histogram using the built-in Excel **Analysis Tool,** then verify that the **Analysis Toolpak** has been installed by selecting **Tools** ⇨ **Add-Ins.** Verify that the **Analysis ToolPak** is selected and click **OK.**

1.   Open the file *ta01_001* from the **IPS CD-ROM.**

2.   Select **Tools** ⇨ **Data Analysis**, select **Histogram** and click **OK.**

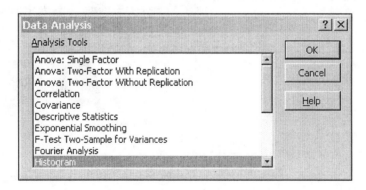

3.   The **Histogram** dialog box requires several inputs. Click inside the **Input Range** box and select the data set without the label.

     How to Select Data: Click on the first cell of the data set, hold down the left mouse button and drag the cursor down to the last cell and release the mouse button. If the dialog box is obstructing the data, there are two methods for accessing it. You can move the dialog box by clicking and dragging the title bar and moving the dialog box to another location. You can also click on the button to the right of the box for **Input**

Range or other input boxes like this. The dialog box will shrink to the size of the input box. When you are done selecting the data range, click on the box to the right of the input box to return to the dialog box and complete the selections.

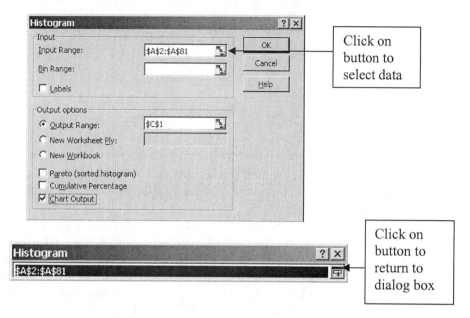

4. Excel will create bins unless you supply your own. The resulting bins will not usually be desirable because the program calculates equal intervals without regard for what looks good in a displayed chart. As an example, we will start by letting Excel create its own bins.

5. Click inside the **Output Range** box under the **Output Options.** Verify that the cursor is inside the box. Otherwise, a previously selected range could be changed. Select a cell to the right of the data set, such as cell B1. This cell will be the upper left cell of the output placed on the worksheet.

6. Select **Chart Output** to create the histogram. Click **OK.**

7. The resulting histogram shows what happens when Excel creates its own bins. The intervals start at 1 with the first number being the only observation in that bin. The interval width is 143.375, not a number that creates an attractive chart. The observations are whole numbers, and it would be appropriate if the bins were whole numbers also. This histogram could be used to preview the data distribution and/or to determine the distribution shape and the approximate number of bins needed. We will use this histogram as a guide to creating your own bins.

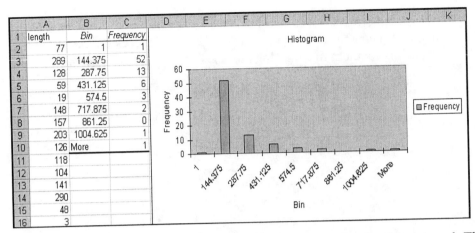

8. **Creating Bins:** Determine the first bin number and an appropriate bin interval. The first bin should contain the first number without skipping a bin. For example, if the bin size is 2 and the first number is 3.5, then the first bin could be 4 or 5, but if you choose 6, you have skipped one whole bin interval. Remember, Excel puts individual values in bins by using the rule "up to and including." It often does not work to select the first number as the first bin. Instead, use the next larger whole number. For example, in this case, choose 100 as the first bin and use bin intervals of 100. Your last bin should include the biggest number. Type the bins yourself, replacing the ones that Excel provided. It is often helpful to sort the data in order to identify the starting and ending bin values (the method is to click anywhere in the data set and click the sort button ![sort icon] in the toolbar). Bin 100 will include the first number, 1, and the largest bin will include the largest number, not including 2631, which is 1148. The **More** category is generally not useful because it gives no specific information about the numbers placed in that category.

9. Input the heading "Bin" in cell B1 and type the bin numbers from 100-1200, with intervals of 100 below the heading. The choice of bins using this method will require some planning to ensure that all numbers are included and the number of bins is appropriate.

Bin
100
200
300
400
500
600
700
800
900
1000
1100
1200

10. Select **Tools** ⇨ **Data Analysis.** Select **Histogram** and click **OK.**

11. The same data will be input into the **Histogram** dialog box as before, but this time, the **Bin Range** will be selected and the **Output Range** should be selected as a cell in the column to the right of the bins. Click **OK.**

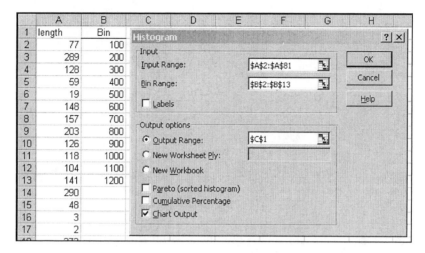

12. The resulting histogram still requires several changes. First, eliminate the **More** category by selecting the histogram chart by clicking on it once, then drag the lower-right blue box (at the bottom of the **Frequency** column) up one line.

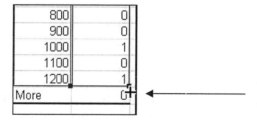

13. Delete the legend by clicking on it once and press **Delete.**

14. Eliminate the gaps between the bars by double-clicking on any bar and selecting the **Options** tab. Change the **Gap width** to "0" and click **OK.**

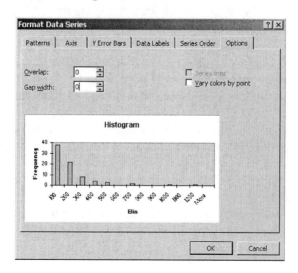

15. Resize the chart and change the title and labels.

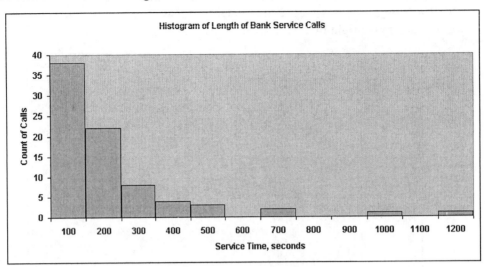

Title: Click inside the box containing the title **Histogram.** Verify that the cursor is at the end of the word "Histogram" and add "of Length of Bank Service Calls".

Bins: Click inside the box containing the label "Bin" and change to "Service Time, seconds".

Frequency: Click inside the box containing the label "Frequency". Click once more inside the box to see the cursor. Change to "Frequency of Calls" or "Count of Calls".

Number of Decimal Places of Axis Values: Double-click on any of the numbers that you want to change, for example 100.00, and select the **Number** tab. Then select the **Number** category on the left and input the desired decimal places, in this case 0. Click **OK.**

Remarks

The distribution of length of service calls shows a strong right skew with most of the calls being in the 100 and 200 second bins. Outliers would be the higher numbers in the 1000 and 1200 second bins. The spread is from 1 second to 1148 seconds if the 2631 second call is left out (spread is identified using exact values, not the bin designations). It was later determined that calls under 10 seconds were actually "hang ups." The employees were trying to lower their service time by deliberately adding short calls.

Excel puts individual values in bins by using the rule "up to and including." For example, all numbers greater than 100 and less than or equal to 200 would be placed in the 200 bin.

The number of bins can affect the shape of a histogram distribution. Too many bins produce too many peaks and gaps between columns, and too few bins produce a couple of large columns with no clear shape. It is worthwhile to think about bin sizes and the number of bins before creating a histogram.

IPS Example 1.7 5th Grade IQ Scores

You have probably heard that the distribution of scores on IQ tests is supposed to be roughly "bell-shaped." Let's look at some actual IQ scores.

Table 1.3 displays the IQ scores of 60 5th grade students chosen at random from one school.

1. Open the file *ta01_003.*

2. Sort the data to help in determining the first and last bins and an appropriate bin interval.

3. There are 60 data points ranging from 81 to 145. If we decide that we want approximately 8 bins, we can determine the bin interval by dividing the total range of data (145 – 81) by the desired number of bins (8). The answer is 8. A more appropriate bin interval might be 10, which would present a more readable chart. The first bin would be 90 to include the lowest number 81. The largest bin would be 150 to include the highest number 145.

Bin
90
100
110
120
130
140
150

4. Select **Tools** ⇨ **Data Analysis.** Select **Histogram** and click **OK.**

5. Select the data set into the **Histogram** dialog box and the **Bin Range.** Select an **Output Range** as a cell in the column to the right of the bins. Click **OK.**

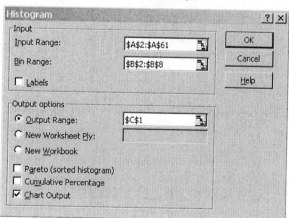

6. Change the titles to reflect this example. Eliminate the **More** category and delete the legend.

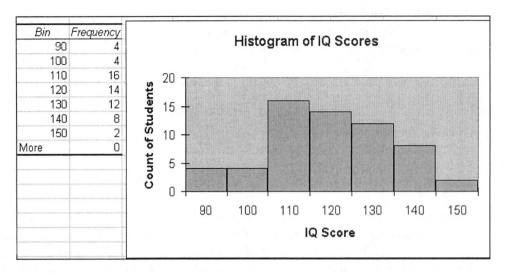

Bin	Frequency
90	4
100	4
110	16
120	14
130	12
140	8
150	2
More	0

Histogram of IQ Scores

7. Occasionally, you may be interested in the percent of observations that fall into each category or bin. There is no automatic way to do that in Excel, but there is a manual method that can be used. Start by typing a new heading "Percent" in the column to the right of the bins.

8. Click in the cell below the **Percent** heading and type "=" and then click on the cell to the left containing the count 4. Type " / 60", which divides that cell by 60, the total number of observations. Press **Enter.**

9. Copy that formula down for the remaining cells.

Frequency	Percent
4	0.066667
4	0.066667
16	0.266667
14	0.233333
12	0.2
8	0.133333
2	

10. Change the format of the cells by selecting the percent cells and clicking on the **Percent** button % in the toolbar.

11. Click once on the histogram chart and then click the **ChartWizard** in the menu. Click **Next** to go to **Step 2.**

12. Click the **Series** tab. Click on the button to the right of the **Values** box and select the newly created percentage values.

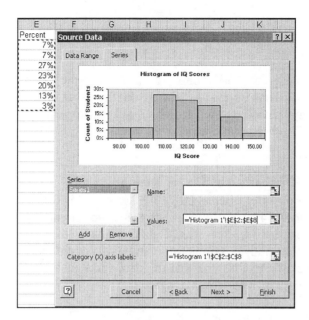

13. Click **Finish.** The histogram now has percentages, instead of counts, for the *y* axis. Change the axis title.

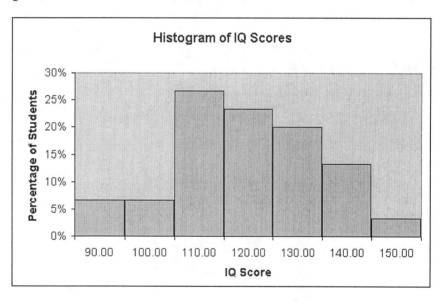

Remarks

The histogram shape looks slightly different from the figure in the *IPS* text because data are placed in bins using different methods in software programs other than Excel. Excel puts the data into the bins by using the "up to and including method". This means that the IQ scores 81, 82, 89, and 90 are placed into the 90 bin, resulting in a count of 4.

The distribution of IQ scores is fairly symmetric, or "bell-shaped," with a single peak in the center. The center is approximately 110–120 IQ and there are no outliers.

Time Plots

A **time plot** of a variable graphs each observation against the time at which that observation was measured. Time is always placed on the horizontal or *x* axis. The variable measured is always placed on the vertical or *y* axis. Connecting the data points with lines helps to define the changes and trends over time.

IPS Example 1.10 Water Discharged by the Mississippi

Table 1.4 lists the volume of water discharged by the Mississippi River into the Gulf of Mexico for each year from 1954 to 2001. The units are cubic kilometers of water. Both graphs in Figure 1.8 describe these data. The histogram shows the distribution of the volume discharged. The histogram is symmetric unimodal (a single peak), with center near 550 cubic kilometers. We might think that the data show just chance year-to-year fluctuation in river level about its long-term average.

We will reproduce the time plot of the same data.

1. Open the file ***ta01_004.*** The data are given as years in the first column and river discharge in cubic kilometers of water in the second column.

2. Select the first and second column of data as a block to create the time plot. Always verify that the *x* variable is in the first column and the *y* variable is in the second column.

	A	B
1	Year	Discharge
2	1954	290
3	1955	420
4	1956	390
5	1957	610
6	1958	550
7	1959	440
8	1960	470
9	1961	600
10	1962	550
11	1963	360
12	1964	390
13	1965	500
14	1966	410
15	1967	460
16	1968	510
17	1969	560
18	1970	540
19	1971	
20	1972	600

3. Click the **ChartWizard** in the menu to access **Step 1** of the **ChartWizard**, and select **XY (Scatter)** for **Chart type** and **Scatter with data points connected by lines** for the **Chart sub-type.** Click **Next.**

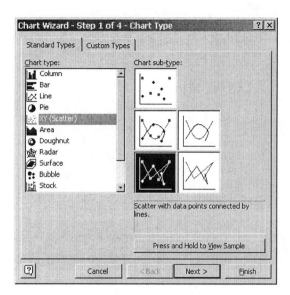

4. **Step 2** of the **ChartWizard** displays a preview of the time plot. Click **Next.**

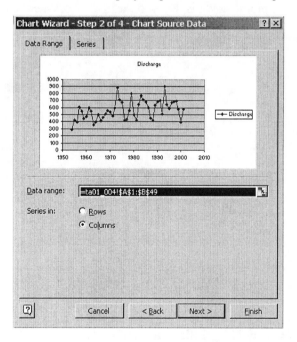

5. In **Step 3,** select the **Titles** tab to input the chart title "Mississippi River Discharge ". Input the **Category (X) axis** label as "Year" and the **Value (Y) axis** label as "River Discharge, cubic kilometers". Select the **Legend** tab and de-select the legend. Click **Next.**

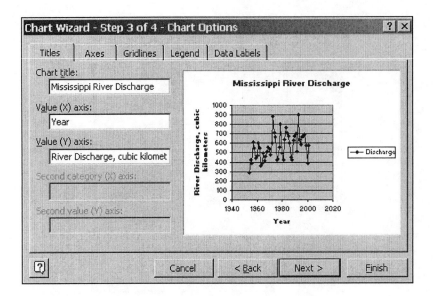

6. **Step 4** of the **ChartWizard** offers you the choice to place the chart **As a new sheet** or embed the chart **As an object** in the current worksheet. Select **As new sheet** and click **Finish**. The time plot is now displayed on its own worksheet entitled **Chart 1.**

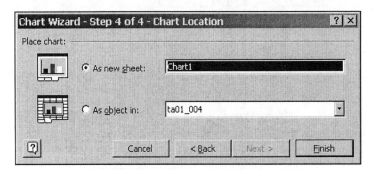

Enhancing the Time Plot

Gridlines:
Gridlines can be removed to make the resulting plot easier to read. Right-click on any of the **gridlines** and select **Clear.** The **gridlines** are removed.

Scaling:
Both the *x* axis and *y* axis scales can be changed to display fewer or more units, as well as changing the maximum and minimum.

1. Right-click on the axis that you want to change and select the options required to change. In this example, right-click on the *y* axis and select **Format Axis.**

2. Click the **Scale** tab and input the **Minimum** as "200" and the **Maximum** as "1000".
    Click **OK.**

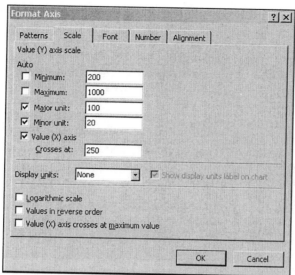

3. Right-click on the *x* axis and make the following changes:

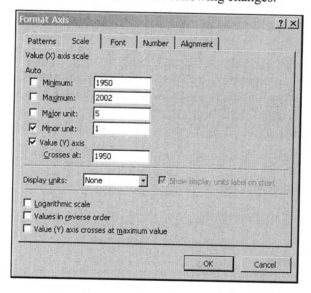

The format of the numbers and the fonts can also be changed using this dialog box. Click
OK when finished.

4. Adding a *trendline* to the time plot will give us a clearer picture of the trend over
    time. This line in Excel is a *best fit* line, which is discussed further in Chapter 2 when
    we create scatterplots and regression analysis. Right-click anywhere on the
    connecting line and select **Add Trendline** and click **OK.** This accepts the default
    straight line.

5. The resulting time plot displays the trends over time in a clear format.

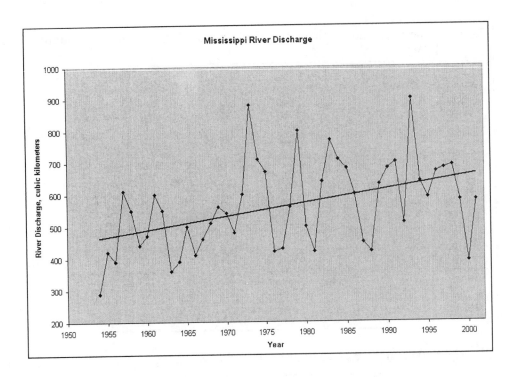

Remarks

The time plot gives us another dimension in analyzing the river discharge data. Although there is clearly variation from year to year, the overall trend over time is clearly increasing. That is, there is a long-term rise in the volume of water discharged due to climate change.

1.2 Describing Distributions with Numbers

It is important to describe data distributions quantitatively in order to be specific about a distribution's **center** and **spread**. The **mean** of a set of n observations is calculated as the average of those observations: $\bar{x} = \dfrac{x_1 + x_2 + \cdots + x_n}{n}$.

Using summation notation: $\bar{x} = \dfrac{1}{n}\sum x_i$, where n is the number of observations and x_i are the values of n observations.

$\sum$ is Greek notation for summation or adding up the values to the right of it.

The bar over the x indicates the mean of all of the x values.

The subcripts identify each individual observation.

The median is defined as the midpoint of the observations. That is, half of the observations are smaller than the median and half are larger than the median. If the number of observations n is an odd number, the median M is defined by the value of the observation located at $(n+1)/2$. If the number of observations is even, the median M is defined as the mean of the two center observations, located by the previous equation.

The spread of a distribution, particularly a non-symmetric or skewed distribution, can be described by **quartiles.** If the observations are ordered from smallest to largest, each quartile represents 25% of the observations. The first quartile (Q_1) represents the median of the observations ordered from the minimum to the overall median M. The second quartile is the overall median M and represents 50% of all observations. The third quartile represents the median of the upper 50% of the observations. **A five-number summary** gives a complete description of the distribution, including the minimum number, Q_1, M (median), Q_3, and the maximum number. A boxplot is a graph of the five-number summary. Side-by-side boxplots are useful to compare several distributions.

The **standard deviation** is a measure of the spread of a distribution. A symmetric distribution is completely described by its center at the mean and its spread defined by multiples of its standard deviation. The **variance** of a set of observations is defined as the average of the squares of the deviations of the observations from their mean. The **variance** of n observations is defined by:

$$s^2 = \frac{(x_1 - \bar{x})^2 + (x_2 - \bar{x})^2 + \cdots + (x_n - \bar{x})^2}{n-1}$$

Or, using summation notation:

$$s^2 = \frac{1}{n-1}\sum(x_i - \bar{x})^2$$

The **standard deviation** s is the square root of the **variance** s^2.

$$s = \sqrt{\frac{1}{n-1}\sum(x_i - \bar{x})^2}$$

The denominator of all of these expressions contains $n - 1$, which is defined as the **degrees of freedom.**

Because the standard deviation measures deviations from the means, it is susceptible to influence by outliers and skewed distributions and, therefore, is not a resistant measure. The standard deviation and mean are more appropriate measures for use with symmetric distributions. Skewed distributions or distributions with outliers are often best represented by the five-number summary.

Descriptive statistics can be obtained from using the built-in tools in Excel.

Generating Descriptive Statistics

IPS Example 1.14 Highway Mileages of Two-Seater Cars

The highway mileages of the 20 gasoline-powered two-seater cars, arranged in order, are

13	15	16	16	17	19	20	22	23	23	23	24	25	25	26	28	28	28	29	32

1. Input the data into a single column on an Excel worksheet.

2. The data are already sorted from low to high and you can observe that the median would lie between the middle two numbers shown above, that is, between two of the 23 values.

3. Select **Tools** ⇨ **Data Analysis** from the menu. Select **Descriptive Statistics** and click **OK.**

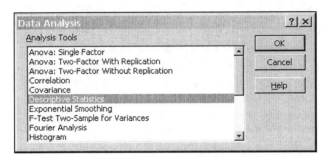

4. Select the entire data set for the **Input Range.** If you typed a label and selected it, select **Labels in First Row.** Click in the box for **Output Range** and click on a blank cell on your worksheet away from the data, representing the upper level of the output region. You can also place the output on a separate worksheet or file. Select **Summary Statistics** and click **OK.**

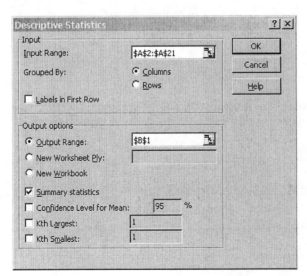

5. The resulting output displays several calculations but *not* the five-number summary. Adjust the column width to display the headings in the first column.

Column1	
Mean	22.6
Standard Error	1.181881
Median	23
Mode	23
Standard Deviation	5.285531
Sample Variance	27.93684
Kurtosis	-0.84927
Skewness	-0.19951
Range	19
Minimum	13
Maximum	32
Sum	452
Count	20

Functions

The mean and median of a data set can be calculated separately (as many of the other values in the **Descriptive Statistics** output) by manually inputting built-in functions in Excel. There are dozens of functions built into Excel. There are two primary methods of inserting functions:

Using the Function Wizard

All of Excel's functions can be accessed by selecting **Insert** ⇨ **Function** from the menu. **Note:** The procedure described in this paragraph, and by the **Paste Function** dialog box shown below, are appropriate for the 2000 version of Excel but may not be appropriate for XP or other versions. Functions are organized by **Function Category** on the left and, within each category, the **Function Names** are listed on the right. When a **Function Name** is selected, it is described at the bottom of the dialog box. If you click **OK,** the **Function Wizard** leads you through the process of inputting the function and performing the necessary calculation. The format for each function is different and described in this method. To start, click in the empty cell where you want the calculation to be performed.

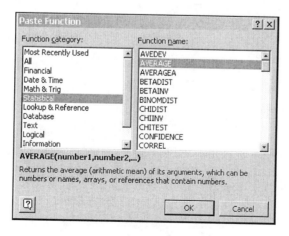

Manual Input of Functions

A function can be inserted into a worksheet by simply typing it into a cell, assuming that you are familiar with the syntax. For example, the mean is calculated by the **AVERAGE** function. To start, click in an empty cell where you want the calculation to be performed. All equations and functions require you to type an "=" sign first, then type the function name (it is not case sensitive) and a left parenthesis. Usually, the data set is selected next (see syntax for each specific function); then type a right parenthesis and press **Enter.**

To manually input the functions for mean and median using the current income data sets, follow the steps outlined below:

1. Input a label to the left or above your calculation to identify its result (in this case, "Mean" and "Median").

2. Click in an empty cell and type "=" and then type the name of the function.

3. Type a left parenthesis and then select the data set for one group, for example, Black Females.

4. **Enter** using the keyboard or type a right parenthesis and then press **Enter.**

5. The value is calculated and displayed in the current cell. The equation for that cell is displayed in **Formula Bar** above the worksheet. Changes can be made if necessary by clicking in the **Formula Bar.**

COUNT	X ✓ =	=AVERAGE(B2:B21)
A	**B**	**C**
1	Mileage	
2	13	
3	15	
4	16	
5	16	
6	17	
7	19	
8	20	
9	22	
10	23	
11	23	
12	23	
13	24	
14	25	
15	25	
16	26	
17	28	
18	28	
19	28	
20	29	
21	32	
22 Mean	=AVERAGE(B2:B21)	

6. To format the answer, click on the cell to be formatted and click the comma button in the toolbar for two decimals, or you can click one of the other format option buttons for either changing to percent or increasing or decreasing the decimal point placement.

% , .00 .0

Copying a Formula

Once input, a formula can be copied horizontally or vertically to perform the same calculation for multiple data sets. Follow the steps outlined below:

1. Click on the cell with the formula to be copied (you can select more than one formula).

2. Move the mouse pointer until a cross hair is visible in the lower right of the cell.

3. Click and drag in the direction that you want to copy the formula (for an additional data set, to the right).

21		32
22	Mean	22.6
23	Median	23
24		

Other functions can be input in the same way to calculate maximum, minimum, count, and dozens of other calculations. See Excel **Help** for further instructions.

Five-Number Summaries and Boxplots

The center of a data set is determined by the mean or the median and the range of values determined by the maximum and minimum. Calculating a five-number summary and graphing the result as a **boxplot**, yields a more complete definition of how a data set is distributed. The five numbers separate the data set into four equal parts called quartiles. Each quartile contains 25% of the data. **Boxplots** are a powerful graphic tool used to compare distributions and are helpful in describing non-symmetric distributions.

Creating a Five-Number Summary Manually

The five-number summary can be created by manually inputting functions for each of the quartiles.

IPS Example 1.14 Highway Mileages of Two-Seater Cars

Use the highway mileages of the 20 gasoline-powered two-seater cars, arranged in order.

13	15	16	16	17	19	20	22	23	23	23	24	25	25	26	28	28	28	29	32

1. Input the data into a single column or copy and paste from the previously examined descriptive statistics example.

2. Below the data and one column to the right, input the following labels and formulas according to the instructions in the previous section by using the **Function Wizard** or by inputting the formulas manually.

5 Number Summary	
Maximum	=MAX(B2:B21)
Q3	=QUARTILE(B2:B21,3)
Median	=MEDIAN(B2:B21)
Q1	=QUARTILE(B2:B21,1)
Minimum	=MIN(B2:B21)

3. The resulting five-number summary is shown below.

5 Number Summary	
Maximum	32.0
Q3	26.5
Median	23.0
Q1	18.5
Minimum	13.0

The STDEV Function

The standard deviation can be calculated by using the **STDEV** function in Excel, either by manual input ("=STDEV(select data)") or by using the **Function Wizard**.

Using the current highway mileage data set, the standard deviation calculation is shown below as 5.29.

Standard Deviation	5.29

Creating a Five-Number Summary and Boxplot

Excel does not have a tool to create a five-number summary or a boxplot. A macro has been programmed by the author of this manual to automate the process. It can be accessed from the IPS Web site http://bcs.whfreeman.com/ips5e/. You can download it as a file to your Desktop or to a folder.

IPS Example 1.16 Creating a Boxplot of the Call Length Data

1. Download the **Create Boxplot** macro from the IPS Web site: http://bcs.whfreeman.com/ips5e/.

2. After downloading the macro file, open it when prompted. You can also double-click on the file to open it from the location where you saved it.

3. The worksheet has a button on it entitled **Create a Boxplot.** When you click on the button, the macro is activated. First we need to have a data set to analyze.

4. Open the length of service call file ***ta01_001.***

5. Copy and paste the data set onto the worksheet where the **Create a Boxplot** macro button is located. Delete the outlier 2631.

6. Click on the **Create Boxplot** button.

7. Select the entire data set. Click **Yes** for the **Identify Outliers** option. If a label is included in the selection, click **Yes** for **First Row Contains Label.** You can also add a label to the chart after it is created. Click **OK.**

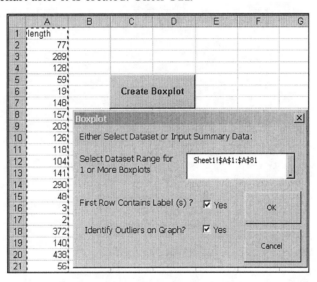

8. The resulting **five-number summary** and **boxplot** is displayed on a new worksheet entitled **Box 1.** Outliers are calculated and listed below the boxplots and graphed with unique markers. The outliers are calculated by the **1.5 IQR rule** (1.5*(Q3-Q1)).

9. The resulting graph can be re-scaled to focus more clearly on the boxplot. Right-click on the *y* axis and select **Format Axis.** Click the **Scale** tab. Change the **Maximum** to "1200" and the **Minimum** to "0". Click **OK.**

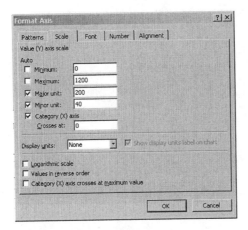

10. Expand the *x* axis title to "Service Call Length" by changing the existing title or by inputting the title into cell B24. The resulting boxplot is show below.

11. A *y* axis label is added by right-clicking once on the graph and selecting **Chart Options.** Type the *y* axis label "Earnings". Click **Finish.**

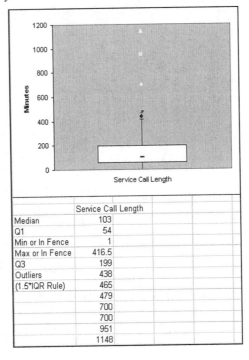

	Service Call Length		
Median	103		
Q1	54		
Min or In Fence	1		
Max or In Fence	416.5		
Q3	199		
Outliers	438		
(1.5*IQR Rule)	465		
	479		
	700		
	700		
	951		
	1148		

Remarks

The boxplot provides a clear graphic representation of data distribution. The boxplot also provides information on whether a data set is symmetric or skewed. The lower line of the box represents the first quartile. The median is shown by the dash and the top of the box represents the third quartile. The minimum value is shown at the bottom and the maximum at the top of the distribution, connected to the box with lines. Outliers are plotted with distinct markers and considered to be outside of the data range. The outliers are determined by the **1.5*IQR rule**. That is, any observations above [Q3 + 1.5*IQR (Q3–Q1)] would be high outliers and any observations below [Q1 – 1.5*IQR] would be low outliers. They are not plotted as either maximum or minimum values. If outliers do exist, the plot whiskers extend to the 1.5*IQR limit, rather than the maximum or minimum.

Creating Side-by-Side Boxplots

IPS Table 1.10 Mileage Comparison

1. Open the file *ta01_010.* The data are given as miles per gallon. Download and open the **Create Boxplot** macro from the IPS web site: http://bcs.whfreeman.com/ips5e/ or open it from the saved location.

2. Copy and paste the data set to the worksheet with the macro button. Separate the data into the four headings as shown below:

Two City	Two Hwy	Mini City	Mini Hwy
17	24	12	19
20	28	21	29
20	28	19	27
17	25	19	28
18	25	16	23
12	20	18	26
11	16	16	23
10	16	18	23
17	23	25	32
60	66	23	31
9	15	20	29
9	13	18	26
15	22	14	22
12	17		
22	28		
16	23		
13	19		
20	26		
20	29		
15	23		
26	32		

3. Click on the macro button and select all of the data, including the headings. Click **Yes** for the **Identify Outliers** option and **Yes** for **First Row Contains Label.** Click **OK.**

4. Click on the chart once and then select **Chart** ⇨ **Chart Options** and click on the **Titles** tab and input "Miles per gallon" as the *y* axis title. The resulting side-by-side boxplot and calculations are shown on the following page.

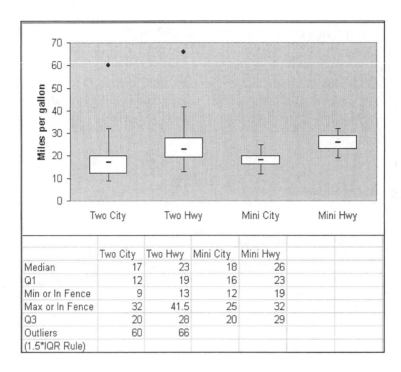

	Two City	Two Hwy	Mini City	Mini Hwy	
Median	17	23	18	26	
Q1	12	19	16	23	
Min or In Fence	9	13	12	19	
Max or In Fence	32	41.5	25	32	
Q3	20	28	20	29	
Outliers	60	66			
(1.5*IQR Rule)					

Remarks

Side-by-side boxplots are a clear graphic comparison of multiple data sets. The highway mileages are higher overall than the city mileages. The minicompact cars have slightly higher median gas mileages than the two-seaters, and their mileages are less variable. The outliers of the Honda Insight are clearly outside the pattern of the two-seater distribution.

Changing the Unit of Measurement

A variable is usually recorded in a specific unit of measurement, such as miles per gallon. However, the units can easily be converted from one unit of measurement to another, such as kilometers per liter. The unit conversion is a linear transformation of the measurements. This means that the original variable x is converted to the new variable x_{new} by using the form of a linear equation.

$$x_{new} = a + bx$$

The constant a shifts all values of x by the same amount. Multiplying by the positive constant b changes the size of the unit of measurement. Some transformations only involve a constant, such as the conversion from miles per gallon to kilometers per liter.

$$x_{new} = 2.35x$$

Other transformations involve a constant, such as the conversion from Fahrenheit to Celsius. Linear transformations do not change the shape of a distribution, but the center and spread will change.

IPS Example 1.20 Unit Transformation

Mary and John both measure the weights of the same five newly hatched pythons. Mary measures in ounces and John uses grams. There are 28.35 grams in an ounce, so each of John's measurements is 28.35 times as large as Mary's measurement of the same python. Here are their results:

Python	1	2	3	4	5
Mary	1.13 oz	1.02 oz	1.23 oz	1.06 oz	1.16 oz
John	32 g	29 g	35 g	30 g	33 g

1. Input the heading row and Mary's data into an Excel worksheet, starting in cell A1. Do not include the units into the cells. If you do, Excel will *not* read these values as numbers. You can put the units in the final column by themselves.

2. Input a new row entitled "New", which will represent the conversion from ounces to grams. Input the formula "=B2*28.35" into cell B3 and press **Enter.** Copy the formula to the right.

	A	B	C	D	E	F
1	Python	1	2	3	4	5
2	Mary	1.13	1.02	1.23	1.06	1.16
3	New	=B2*28.35				

3. The resulting row should equal John's measurement in grams. In comparing the values, we see that the result is equal to John's measurements if the answer is rounded to the nearest whole gram.

	A	B	C	D	E	F
1	Python	1	2	3	4	5
2	Mary	1.13	1.02	1.23	1.06	1.16
3	New	32.0355	28.917	34.8705	30.051	32.886
4						

4. Select the calculated values in the third row and click the decrease decimal button ![decrease decimal icon] in the toolbar until only a whole number is displayed.

5. Calculate the values of the means (use the average function) separately for Mary and John's measurements.

	A	B	C	D	E	F	G	H
1	Python	1	2	3	4	5	Means	
2	Mary	1.13	1.02	1.23	1.06	1.16	1.12	oz
3	John	32	29	35	30	33	31.8	g

Remarks

The mean for John's measurements is also the same ratio, 28.35 times Mary's mean. The spread or standard deviation would also reflect the same ratio.

If a constant a is added to each observation, the value a would be added to measures of center and to the quartiles and other percentiles, but it does not change the measure of spread.

1.3 Density Curves and Normal Distributions

Normal distributions are defined as perfectly symmetric, bell-shaped curves described in the text as density curves. The normal distribution has certain qualities that allow calculations based on standardized values of the standard deviation.

If a variable x has a normal distribution $N(\mu, \sigma)$ with a mean μ and a standard deviation σ, then the standardized variable $z = \dfrac{x - \mu}{\sigma}$ also has a standard normal distribution.

Normal distribution calculations determine the probability that a distribution would have a given value at least that high or higher, or at least that low or lower. The calculations can also determine a specific value for a given probability or proportion. These methods are described in the text.

Normal Distribution Calculations

Finding a Proportion When Given a Value

Excel functions can calculate normal distribution values. The manual method using Table A, as defined in the text, is the basis of these calculations.

IPS Example 1.25 Normal Distribution Calculations

The National Collegiate Athletic Association (NCAA) requires Division I athletes to score at least 820 on the combined mathematics and verbal parts of the SAT exam in order to compete in their first college year. The scores of the 1.4 million students in the class of 2003 who took the SATs were approximately normal with mean 1026 and standard deviation 209. What proportion of all students had SAT scores of at least 820?

1.   Open an empty worksheet in Excel.

2. Type in the given values and their headings. We are interested in the proportion or probability of having a value as great as or greater than 820. However, because of how the normal distribution table (Table A in the text) is constructed, the area to the left of a specific value is always calculated. If an area to the right of a specific value is desired, we will have to subtract the value from 1.

Ex. 1.25	
Pop Mean	1026
Pop SDev	209
x	820
Z-Score	
P < 820	
P > 820	

3. Click in the cell to the right of the **Z-Score** to input the function to calculate the standardized *z* value of a specific *x* value. Input "= STANDARDIZE (820,1026,209)" and press **Enter.** The arguments of the function in order are the *x* value, the population mean, and the population standard deviation. Instead of typing in the specific values, you can click on the cells containing those values.

4. Click in the cell to the right of the **P < 820** cell to input the function to calculate the probability of an *x* value being as high or higher than 820. Input "= NORMSDIST(", click on the *z* score in the cell above (on my spreadsheet, located in cell J7), and press **Enter.**

Ex. 1.25	
Pop Mean	1026
Pop SDev	209
x	820
Z-Score	=STANDARDIZE(820,1026,209)
P < 820	=NORMSDIST(J7)

Z-Score	-0.9856
P < 820	0.16215
P > 820	

5. Click in the cell to the right of the **P > 820** cell and input "= 1– " and click on the 0.16215 in the previous cell and press **Enter.**

Z-Score	-0.9856
P < 820	0.16215
P > 820	0.83785

Remarks

The resulting proportion equals 0.8379, or approximately 84%. That is the proportion of all SAT-takers who would be NCAA qualifiers.

Finding a Proportion Between Two Values

Sometimes it is desired to calculate the proportion between two values. The solution uses the function in the previous example, however, it is important to understand the process for calculating the in-between proportion so that there is no confusion.

If we want to calculate the proportion between values *a* and *b*, you would calculate the area to the left of *b* and then subtract the area to the left of *a*. This process is pictured below.

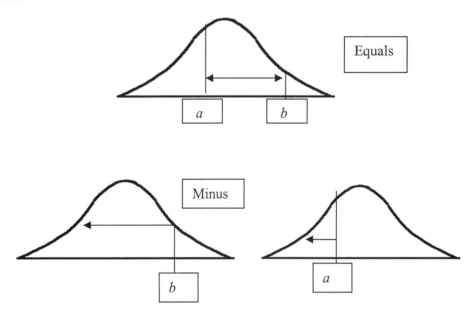

IPS Example 1.26 Normal Distribution Calculations for In-Between Values

The NCAA considers a student a "partial qualifier," eligible to practice and receive an athletic scholarship, not to compete, if the combined SAT score is at least 720. What proportion of all students who take the SAT would be partial qualifiers? That is, what proportion have scores between 720 and 820?

1. Open an empty worksheet in Excel.

2. Input the headings and values from the previous example. Copy and paste the *z* score for 820 and the *P* value < 820.

3. Another *z* score will be calculated for 720 with its corresponding *P* value.

Ex. 1.26	
Pop Mean	1026
Pop SDev	209
x1	820
x2	720
Z-Score	-0.98565
P < 820	0.16215
Z-Score	
P < 720	

4. Input the **Z-Score** function for 720 as "= STANDARDIZE (720,1026,209)" and press **Enter.**

5. Input the **P < 720** function as "= NORMSDIST(", click on the *z* score in the cell above, and press **Enter.**

Ex. 1.26	
Pop Mean	1026
Pop SDev	209
x1	820
x2	720
Z-Score	-0.98565
P < 820	0.16215
Z-Score	-1.46411
P < 720	0.07158

6. The in-between proportion is the proportion < 820, which is 0.16215, minus the proportion < 720, which is 0.07158. Input the appropriate equation and calculate the result.

Ex. 1.26	
Pop Mean	1026
Pop SDev	209
x1	820
x2	720
Z-Score	-0.98565
P < 820	0.16215
Z-Score	-1.46411
P < 720	0.07158
In-Between	0.09057

Remarks

The proportion between 720 and 820 is calculated to be approximately 9% of all students who take the SAT.

Inverse Normal Calculations
or Finding a Value When Given a Proportion

IPS Example 1.30 "Backward" Normal Calculations

Scores on the SAT verbal test in recent years approximately follow the N (505,110) distribution. How high must a student score in order to place in the top 10% of all students taking the SAT?

1. Input the given data into an Excel worksheet.

Ex. 1.30	
Pop Mean	505
Pop SDev	110
%	10%
Z-Score	
x value	

2. Calculate the z value by inputting the formula "= NORMSINV(0.9)" and press **Enter.** This function calculates the z value for a proportion of 0.9 or 90%. The reason that we use 90% is because the standardized table (Table A) is constructed always giving the proportion that is less than a given value. When you want to find the value corresponding to an upper percentage, you subtract that percentage from 100% to find the appropriate z value.

3. Input the x value formula corresponding to $x = \mu + z\sigma$. The formulas are shown below.

Ex. 1.30	
Pop Mean	505
Pop SDev	110
%	0.1
Z-Score	=NORMSINV(0.9)
x value	=B3+B7*B4

Remarks

The resulting x value is 645.97, or a student needs to score at least 646 to place in the top 10%.

Ex. 1.30	
Pop Mean	505
Pop SDev	110
%	10%
Z-Score	1.28155
x value	645.971

Assessing the Normality of Data

A data set can be analyzed by creating a histogram and viewing the results, by comparing the mean and the median, or by constructing a **normal quantile plot.** A **normal quantile plot** calculates and graphs standardized values, or *z* values, for each data point. It is important to examine the normality of a data set prior to using statistical procedures that depend on that. Excel does not have a tool to create a **normal quantile plot.** A macro has been programmed by the author of this manual to automate the process. It can be accessed from the IPS Web site http://bcs.whfreeman.com/ips5e/. You can download it as a file to your Desktop or to a folder.

Creating a Normal Quantile Plot

IPS Example 1.31 Normal Quantile Plot of Breaking Strength

1. Open the file *eg01_009.* Download and open the **Create Normal Quantile Plot** macro from the IPS Web site: http://bcs.whfreeman.com/ips5e/ or open it from the saved location.

2. Copy and paste the data set to the worksheet with the macro button.

3. Click on the **Create Normal Quantile Plot** button.

4. Select the entire data set. Input the **Y-Axis Label** as "Breaking Strength, pounds" and click **OK.**

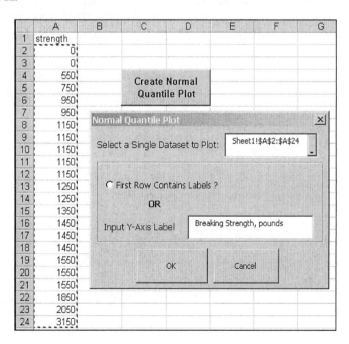

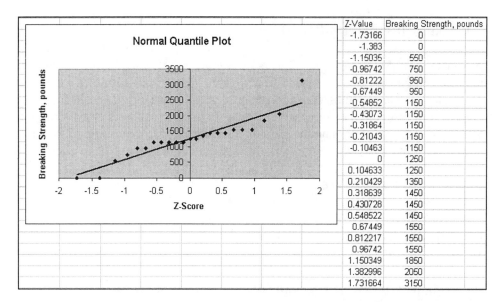

Z-Value	Breaking Strength, pounds
-1.73166	0
-1.383	0
-1.15035	550
-0.96742	750
-0.81222	950
-0.67449	950
-0.54852	1150
-0.43073	1150
-0.31864	1150
-0.21043	1150
-0.10463	1150
0	1250
0.104633	1250
0.210429	1350
0.318639	1450
0.430728	1450
0.548522	1450
0.67449	1550
0.812217	1550
0.96742	1550
1.150349	1850
1.382996	2050
1.731664	3150

5. The resulting plot can be rescaled and enhanced to focus on the shape of the plot. The gridlines can be cleared by right-clicking anywhere on them and selecting **Clear.**

Remarks

How close the points follow the straight line is an indication of how normal the distribution is. Any points that stray significantly from this line are considered to be outliers. Most of the points in this data set lie fairly close to the line. Several data points have the same breaking strength, caused by limited precision of the measurements. A high outlier at 3150 pounds is significantly higher than the line and the rest of the data points. The low outliers at strength of zero are suspicious. It is important to look at overall patterns and clear departures from normality in analyzing the normal quantile plot.

Chapter 2

Looking at Data—Relationships

It is often desirable to study the relationship between two variables measured on the same individuals to see if there is a connection or **association** between the variables. A response variable measures an outcome of a study and is known as the dependent variable y, plotted on the vertical y axis. An explanatory variable explains or affects a change in the response variable and is known as the independent variable x, plotted on the horizontal x axis. These variables model the Cartesian coordinate system shown below with data points graphed with positive values of both variables

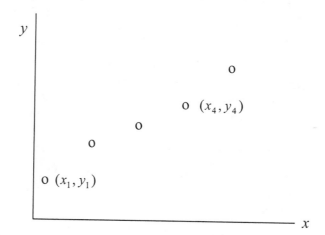

Each data point has both an x and y value denoted by the pair notation (x_i, y_i) where i is an index incrementing each data point (1, 2, 3,…).

2.1 Scatterplots

A plot of observed quantitative x and y data points measured on the same individuals is called a **scatterplot.** x values are plotted on the horizontal axis and y values are plotted on the vertical axis. The **scatterplot** is the useful for displaying the relationhip between two quantitative variables.

IPS Example 2.3 SAT Scores

More than a million high school seniors take the SAT college entrance examination each year. We sometimes see the states "rated" by the average SAT scores of their seniors. For example, Illinois students average 1179 on the SAT, which looks better than the 1038 average of Massachusetts's students. This makes little sense, because average SAT score is largely explained by what percent of a state's students take the SAT. The scatterplot in Figure 2.1 allows us to see how the mean SAT score in each state is related to the percent of that state's high school seniors who take the SAT.

Creating a Scatterplot

1. Open the file *eg02_003*.

2. Decide which variable is *x* and which is *y,* if not stated explicitly in the problem (see step 2 in the previous method). In this case, it is stated that average SAT score is explained by percent of students taking the SAT. Therefore, the explanatory *x* variable is the percent taking SAT and the response variable *y* is the average SAT score.

3. Verify that the *x* variable is in the first column and the *y* variable is in the second column. This order is required to correctly plot an XY Scatterplot in Excel. In this case, the *x* variable is in column B and the mean SAT total is in column E. The variables are in the correct order, but they are not adjacent.

4. Select the cells B1:B52, hold the **Control** key (on the keyboard) down while you select cells E1:E52. Click the **ChartWizard** 📊 in the toolbar to access **Step 1.** Select **XY (Scatter)** for **Chart type** on the left and select the top **Chart sub-type** on the right. Click **Next**.

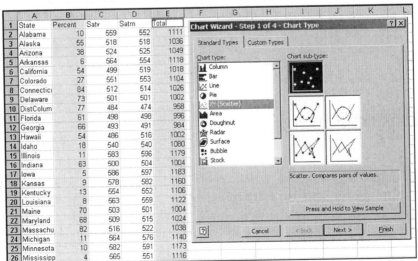

Note: A line graph is appropriate for time plots but will not work for scatterplots.

5. **Step 2** of the **ChartWizard** previews the chart and allows changes to the **Data Range.** Click **Next.**

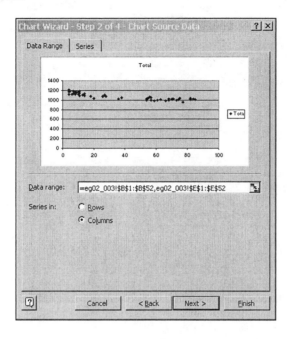

6. **Step 3** allows you to input your chart title and axes labels as well as other options. Click the **Titles** tab and input "Percent Taking SAT" for the **Value (X) Axis** and "Mean SAT Total Score" for the **Value (Y) Axis.**

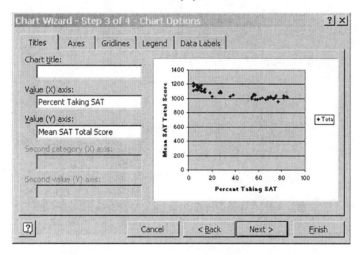

7. Click the **Legend** tab and deselect the **Show Legend** option. The legend will be appropriate for graphing more than one series. Click **Next.**

8. **Step 4** of the **ChartWizard** gives you the choice to place the chart **As a new sheet** or embed the chart **As an object** in the current worksheet. Accept the default **As object in:** by clicking **Finish.**

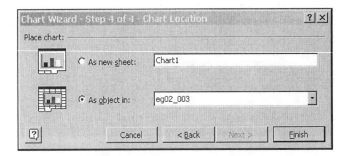

The scatterplot is now embedded on your worksheet next to your data. Resize and change fonts as appropriate.

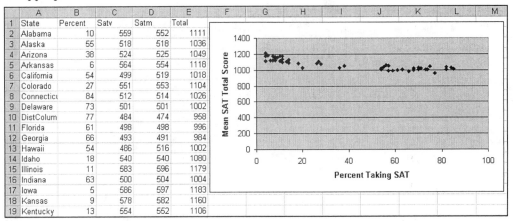

Enhancing a Scatterplot

A scatterplot can be enhanced as any other chart using some of the options provided in the **ChartWizard**. Access the existing chart and its options by performing the following steps:

1. Select the chart (by clicking once within its borders).
2. Click the **ChartWizard** 📊 in the toolbar again.
3. Change or add to the options you originally selected.

Note: If you don't select the chart first, you will not be entering the **ChartWizard** with the data input and options first selected.

Changing the Scale

Excel has a default range from 0 to 100%, which in some cases can leave large areas of blank space on a chart. The scales can be changed as follows:

1. Double-click on the axis that you want to change (*x* or *y*) to produce the **Format Axis** dialog box. The alternate method is to right-click on the axis and select **Format Axis.**

2. Click the **Scale** tab.
3. Change the **Maximum, Minimum, Major,** or **Minor** units.
4. For our previous example, a minimum of 800 will focus on the data points.

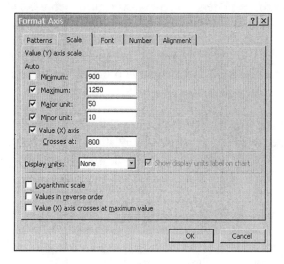

Removing Gridlines

If desired, the gridlines can be removed by right-clicking on any of the gridlines and selecting **Clear.**

Changing the Titles and Labels

To change an existing title or label on a chart, click once to place the cursor inside the label box and add or delete the existing text. If a title doesn't exist, right-click on the chart, select **Chart Options,** and input the titles. Click **OK** to accept the changes.

Our enhanced scatterplot for the previous example is shown below.

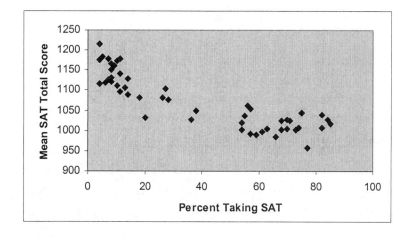

Remarks

The variables in this scatterplot are negatively associated, that is, states that have a higher percent of students taking the SAT tend to have lower mean scores. There appears to be two distinct clusters of states. In one cluster, less than half of the seniors take the SAT with high mean scores. In the other cluster, more than half of the seniors take the SAT with the mean scores being low. Clusters can indicate another variable having an impact. In this case, that variable is the ACT test, which is preferred in the states shown on the left.

The data in the first cluster (under 50%) roughly follows a straight line. There is a moderate amount of scatter resulting in a moderately strong linear relationship. The data in the over 50% cluster shows a weaker relationship between the variables.

Adding Categorical Variables to Scatterplots

A categorical variable is added to a scatterplot by using a different plot color or symbol for each category. This method is demonstrated using data from the previous exercise.

IPS Example 2.3 SAT Scores—Categories

1. Copy and paste data corresponding to the Northeast states to an area to the right of the original data set. Copy and paste data corresponding to the Midwest states below it as shown below.

Northeast	Percent	Satv	Satm	Total
Connecticut	84	512	514	1026
Delaware	73	501	501	1002
Maine	70	503	501	1004
Massachusetts	82	516	522	1038
NewHampshire	75	522	521	1043
NewJersey	85	501	515	1016
NewYork	82	496	510	1006
Pennsylvania	73	500	502	1002
RhodeIsland	74	502	504	1006
Vermont	70	515	512	1027

Midwest	Percent	Satv	Satm	Total
Illinois	11	583	596	1179
Indiana	63	500	504	1004
Iowa	5	586	597	1183
Kansas	9	578	582	1160
Michigan	11	564	576	1140
Minnesota	10	582	591	1173
Missouri	8	582	583	1165
Nebraska	8	573	578	1151
NorthDakota	4	602	613	1215
Ohio	28	536	541	1077
SouthDakota	4	588	588	1176
Wisconsin	7	585	594	1179

2. Select the Northeast percent data and the mean total and create a scatterplot as in the previous exercise (Click the **ChartWizard** 📊 in the toolbar to access **Step 1.** Select **XY (Scatter)** for **Chart type** on the left and select the top **Chart sub-type** on the right. Click **Next.**

3. Click the **Series** tab. Click in the **Name:** box, delete the contents, and type "NE" for Northeast.

4. Click the **Add** button. Click in the **Name:** box and type "MW" for Midwest.

5. Click in the **X Values:** box and select ONLY the *x* values of the Midwest series.

6. Click in the **Y Values:** box and select ONLY the *y* values of the Midwest series.

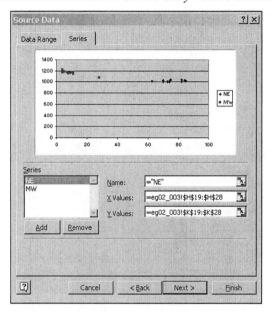

7. Click **Next**. Click the **Legends** tab and reselect **Show Legend.** Click **Finish.** The new scatterplot shows both series.

8. Rescale the resulting graph and clear the gridlines. The legend distinguishes the two groups. The symbols can be changed by right-clicking on the symbol and selecting **Format Data Series**. Change the size, color, and type of symbol.

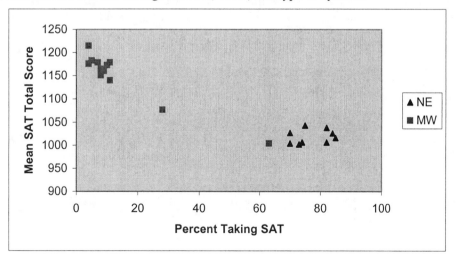

Remarks

The Northeast states are primarily in the above 50% cluster, with the exception of Indiana. The Midwest states are primarily in the lower 50% cluster. The ACT test is preferred in the Midwest states.

2.2 Correlation

A scatterplot investigates the relationship between two quantitative variables and displays information about the form, direction, and strength of the relationship. A numerical measure that quantifies the direction and strength of the linear relationship between two quantitative variables is called the **correlation.** If there are x and y data for n number of points, the first point can be defined as (x_1, y_1), the second as (x_2, y_2), etc., up through (x_i, y_i) where i is an index incrementing each data point $(1, 2, 3, \ldots)$. The **correlation** is written as r and defined by the following equation:

$$r = \frac{1}{n-1} \sum \left(\frac{x_i - \overline{x}}{s_x} \right) \left(\frac{y_i - \overline{y}}{s_y} \right)$$

Where $\overline{x}$ and $\overline{y}$ are the means, and s_x and s_y are the standard deviations of the x and y data. The $\sum$ sign indicates that the terms are summed for all n points.

The CORREL Function

The most efficient way to calculate the correlation coefficient is by using the **CORREL** function in Excel either by manual input as shown on the next page or by using the **Function Wizard.** This function requires that the number of x values must be the same as the number of y values.

IPS Example 2.7 Correlation Example Using SAT Data

1. Open the file *eg02_003.*

2. Click on an empty cell below the x axis data and type the word "Correlation" as a label.

3. Click on the empty cell to the right of the label "Correlation" and type "=CORREL(". Select the x data (Percent), type a comma, and then select the y data (Total). Type a right parenthesis ") " and press **Enter.**

52	Wyoming	11	548	549	1097
53					
54	Correlation	=CORREL(B2:B52,E2:E52)			

51	Wisconsin	7
52	Wyoming	11
53		
54	Correlation	-0.87688

4. The value of the correlation coefficient will be displayed to several decimal points. The value can be reformatted to two or three decimal places as desired. Note that rounding back to two decimal places may round the correlation value to 1, which may not be quite accurate. A value of 0.995 would be more appropriate than 1, which implies a perfect correlation.

Remarks

A value of –0.877 correlation indicates a negative association of moderate high strength.

2.3   Least-Squares Regression

If a scatterplot demonstrates a linear relationship, it would be helpful to summarize this relationship by drawing a line through the data points. A **regression line** is a straight line that describes how the response variable *y* changes as the explanatory variable *x* changes. Because there is more than one way to draw this line, it is appropriate to use a mathematical method that allows accurate predictions. The most common method is the **least-squares** method. This method makes the sum of the squares of the vertical distances of the data points from the line as small as possible.

The **least-squares regression line** is of the form:

$$\hat{y} = a + bx, \text{ where the } \textbf{slope} \quad b = r\frac{s_y}{s_x} \quad \text{ and the } \textbf{\textit{y} intercept} \text{ is } a$$

This equation is in the same form as $y = mx + b$, the equation of a straight line. The slope is always the number that is multiplied by *x,* and the *y* intercept is always the number by itself in the equation. The $\hat{y}$ indicates that the line gives a predicted response for an *x* value. This value is usually different than the observed response *y*.

Fitting a Least-Squares Regression Line to Data

IPS Example 2.9 Fidgeting and Weight Gain

Does fidgeting keep you slim? Some people don't gain weight even when they overeat. Perhaps fidgeting and other "nonexercise activity" (NEA) explains why. Researchers deliberately overfed 16 healthy young adults for 8 weeks. They measured fat gain (in kilograms) and, as an explanatory variable, increase in energy use (in calories) from activity other than deliberate exercise–fidgeting, daily living, and the like. The body might spontaneously increase nonexercise activity when fed more.

1. Open the file *eg02_009.*

2. Verify that the data are in the appropriate *xy* order for a scatterplot. The explanatory variable *x* in this example is nonexercise activity, which is shown in the first column; therefore, the data are in the correct order.

3. Graph the data by first selecting the columns together as a range and then click the **ChartWizard** 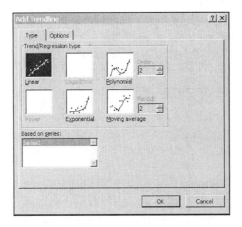 and select the **(XY) Scatter** plot. Click **Next** twice.

4. In **Step 3,** input appropriate labels for the axes. Click **Finish.**

5. If the *y* axis is not on the left side of the scatterplot, right-click on the *x* axis and select **Format Axis** and click the **Scale** tab. Input "–200" for **Value (Y) axis Crosses at.** Click **OK.**

6. Right-click on any data point and select **Add Trendline.** There are several choices for types of lines. Choose the upper left one depicting a straight line and click **OK.**

7. Click the **Options** tab and then select **Display Equation** and **Display R-squared** on the chart. Click **OK.**

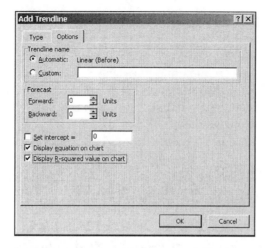

8. The resulting scatterplot displays the line, the equation of that line, and the R^2 value, which is the correlation coefficient squared. The equation and R^2 value are in a graphic box that can be moved anywhere on the graph and resized.

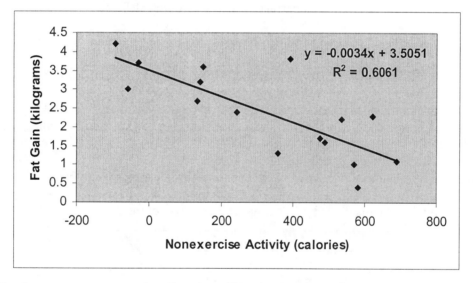

9. The least-squares regression line (trendline in Excel) is $\hat{y} = -0.0034x + 3.5051$, which defines a slope of -0.0034 and a y intercept of 3.5051. The slope indicates that for every one calorie of nonexercise activity, the fat gain in kilograms decreases by 0.0034.

10. Calculate the correlation, as defined in the previous activity. The correlation $r = -0.7785$. The R^2 of 0.61 indicates that approximately 61% of the data points are following the line with almost 40% scatter.

Remarks

The regression line and R^2 value gives us insight into the relationship between these variables. The moderately high negative correlation indicates that as nonexercise activity increases, people gain less fat.

Calculating Predicted Values Using the Regression Line

One purpose of fitting a regression line to a dataset is to be able to predict values either inside (interpolation) or outside (extrapolation) of the given x and y values. If the regression line is an accurate representation of the linear relationship, this same equation can be used to make the predictions.

Manual Calculation of Predicted Values

IPS Example 2.10 Nonexercise Activity and Weight Gain

1. Open the file *eg02_009* used in the previous exercise. Skip to step 4 if you have already graphed the scatterplot with a regression line.

2. Graph this dataset using a **Scatterplot** with appropriate labels.

3. Add a regression line **(trendline)** and its corresponding equation.

4. Add a column to the right of the dataset with a title of **Predicted Values.**

5. In the first cell below the title, type the regression equation "–0.0034* " and then click on the x value for that row (–94). Finish typing the equation by adding "+3.505" and press **Enter.**

6. The calculated value is 3.8246 and represents the equation solved for $x = -94$.

7. Copy the formula down by dragging the lower right of the formula cell.

C2		=	=-0.0034*A2+3.505	
	A	B	C	D
1	nea	fat	Predicted Values	
2	-94	4.2	3.8246	
3	-57	3	3.6988	
4	-29	3.7	3.6036	
5	135	2.7		
6	143	3.2		

Calculating Predicted Values for a Specific x Value

1. Follow the above procedure and then copy the formula down to an empty row below the dataset **OR** input the equation as shown above in an empty row below the dataset.

2. Type any *x* value into the *x* column for that row and press **Enter.** The resulting predicted *y* value will be calculated. In this example, the fat gain for 300 calories of nonexercise acitivity is predicted to be 2.485 kilograms. This is the *y* value on the regression line for the specific *x* value of 300.

17	690	1.1	1.159
18	300		2.485

Remarks

The accuracy of predictions from a regression line (trendline in Excel) depends on the strength of the relationship and how much the data are scattered about the line.

Calculating *x* Values for a Specific *y* Value–Using Goal Seek

1. Follow the above procedure and then copy the formula down to an empty row below the dataset **OR** input the equation as shown above in an empty row below the dataset.

2. Click on the cell with the copied formula to make sure only that cell is selected.

3. Select **Tools** ⇨ **Goal Seek** from the menu.

4. The **Set cell** box contains the cell reference for the formula cell. Click in the **To value** box and type the *y* value that you want to solve the equation for. In this example, input "2.5".

5. Click in the **By changing** box and then click on the *x* cell for that row. Click **OK.**

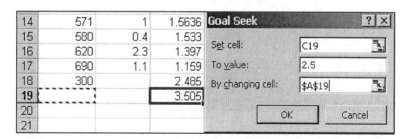

6. The **Goal Seek** tool solves the equation for the *x* value that results in a *y* value of 2.5. Click **OK.** This process can be repeated for any other predictions.

18	300	2.485
19	295.58824	2.5

Remarks

Interpolation, or calculating *x* or *y* values from the regression equation provides fairly accurate results if the correlation is high. However, extrapolation beyond a data set may not prove trustworthy.

Residuals

A **residual** is the difference between an observed value of the response variable y and the y value predicted from the regression line.

$$\textbf{Residual} = \text{observed } y - \text{predicted } y = y - \hat{y}$$

A **residual** represents the vertical distance from the actual y data point to the y value located on the regression line for a particular x value. A **residual plot** is a scatterplot of the regression residuals against the explanatory variable x. Residual plots are used to assess the fit of the regression line and identify lurking variables.

Manual Calculation of Residuals

IPS Example 2.15 Residuals for Fat Gain

Example 2.9 describes measurements on 16 young people who volunteered to overeat for 8 weeks. Those whose nonexercise activity (NEA) spontaneously rose substantially gained less fat than the others. The data were plotted with a least-squares regression line in the previous exercise.

1. Open the file *eg02_009* used in the previous exercise. Skip to **Step 4** if you have already graphed the scatterplot with a regression line.

2. Graph this dataset using a **Scatterplot** with appropriate labels.

3. Add a regression line **(trendline)** and its corresponding equation.

4. Add a column to the right of the dataset with a title of **Predicted Values.**

5. In the first cell below the title, type the regression equation "–0.0034* " and then click on the x value for that row (–94). Finish typing the equation by adding "+3.505" and press **Enter.**

6. Copy the formula down by dragging the lower right of the formula cell.

7. Add a column to the right of **Predicted Values** with the title **Residuals.**

8. In the first cell below the title, type the residual equation (observed – predicted y values). Start with "=" and then click on the observed y value for that row (actual data point 4.2). Type a minus sign and then click on the predicted y value for that row (3.8246) and press **Enter.**

9. The calculated value of the first residual is 0.3754.

10. Copy the formula down by dragging the lower right of the formula cell. The worksheet should look like the one shown below.

	A	B	C	D
1	nea	fat	Predicted Values	Residuals
2	-94	4.2	3.8246	0.3754
3	-57	3	3.6988	-0.6988
4	-29	3.7	3.6036	0.0964
5	135	2.7	3.046	-0.346
6	143	3.2	3.0188	0.1812
7	151	3.6	2.9916	0.6084
8	245	2.4	2.672	-0.272
9	355	1.3	2.298	-0.998
10	392	3.8	2.1722	1.6278
11	473	1.7	1.8968	-0.1968
12	486	1.6	1.8526	-0.2526
13	535	2.2	1.686	0.514
14	571	1	1.5636	-0.5636
15	580	0.4	1.533	-1.133
16	620	2.3	1.397	0.903
17	690	1.1	1.159	-0.059

11. Make a **Scatterplot** of the **Residuals** by selecting the *x* values and then holding the **Control** key (on the keyboard) down while selecting the **Residuals**. Click the **ChartWizard** and select **Scatterplot**. Input appropriate labels and title.

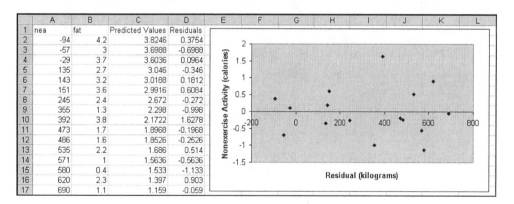

Remarks

The residual plot is scattered. A fairly equal number of points have positive and negative residuals. There is no obvious evidence of other contributing factors.

Outliers and Influential Observations

Detailed examination of a data set's scatterplot and residual plot can lead to discoveries of outliers and influential observations and even lurking variables. A data set is examined to illustrate this point.

IPS Example 2.17 Diabetes Measurements

People with diabetes must manage their blood sugar levels carefully. They measure their fasting plasma glucose (FPG) several times a day with a glucose meter. Another measurement, made at regular medical checkups, is called HbA. This is roughly the percent of red blood cells that have a glucose molecule attached. It measures average exposure to glucose over a period of several months. Here are data on both HbA and FPG for 18 diabetics five months after they had completed a diabetes education class.

1. Open the file *ta02_005*.

2. Graph this dataset using a scatterplot using HbA levels as the *x* variable and the FPG levels as the *y* variable. Input appropriate labels.

3. Add a regression line **(trendline)** and its corresponding equation.

4. Adjust the *x* axis scale to have a minimum of 4 percent and a maximum of 20 percent.

5. Calculate the correlation.

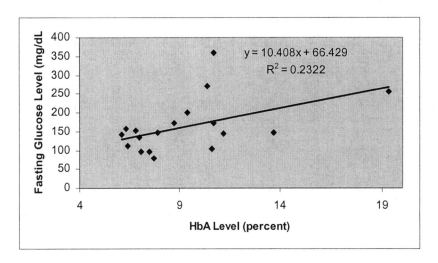

Remarks

The data are fairly scattered with a fairly weak positive correlation of 0.48 and R^2 of 0.23. There are two unusual observations that can be classified as possible outliers. Subject 15 (10.7, 359) displays an unusually high fasting glucose level with a large vertical distance from the line representing a large positive residual. Subject 18 (19.3, 255) displays a high HbA level along the *x* axis and very close to the line.

Outliers like Subject 18 can have a strong influence on the regression line. We will remove the outliers to investigate how **influential** they are, that is, how much they change the resulting calculations for the least-squares regression line. The outliers will be removed individually to determine the effect of each data point.

6. Copy and paste the data set to an area below the original set. This will prevent the original set from changing on the scatterplot.

7. Delete the observation for Subject 15 by deleting the row or deleting the data.

8. Click on the original scatterplot once to select it and then click back on the **ChartWizard** in the toolbar. Click **Next.**

9. Click the **Series** tab.

10. Click inside the **Name** box, delete the contents, and input "All Data" for the legend label.

11. Click **Add** to add a new series without Subject 15.

12. Click inside the **Name** box and input "W/O Subject 15" for the legend label.

13. Click inside the **X Values** box and select only the HbA levels for the copy and pasted data set without Subject 15.

14. Click inside the **Y Values** box and select only the FPG levels for the copy and pasted data set without Subject 15. Click **Finish.**

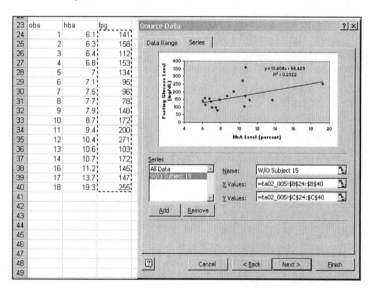

15. The resulting scatterplot displays the new copy and pasted data set overlaying the original data set with the exception of Subject 15. Right-click on any of the overlayed data points and select **Insert Trendline.** Select the **Options** tab and then select **Display Equation** and **Display** R^2.

16. The new trendline and equation can be formatted with a new color to distinguish it from the other trendline representing all data points. In addition, the box containing the new equation can be moved around to enhance the plot.

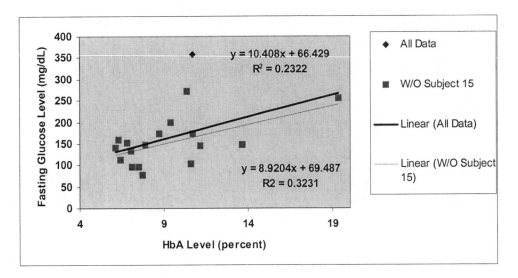

Note: Verify that the legend is displayed. If it is not, right-click on the scatterplot and select **Chart Options.** Then click the **Legend** tab and select **Show Legend.**

Remarks

Removing Subject 15 pulls the regression line up somewhat and increases the R^2 to 0.32. We will now add the data set with Subject 18 removed.

17. Copy and paste the data set to an area below the previous pasted set. This will prevent the original set from changing on the scatterplot.

18. Delete the observation for Subject 18 by deleting the row or deleting the data.

19. Click on the original scatterplot once to select it and then click back on the **ChartWizard** [icon] in the toolbar. Click **Next.**

20. Click the **Series** tab.

21. Click **Add** to add a new series without Subject 18.

22. Click inside the **Name** box and input "W/O Subject 18" for the legend label.

23. Click inside the **X Values** box and select only the HbA levels for the copy and pasted data set without Subject 18.

24. Click inside the **Y Values** box and select only the FPG levels for the copy and pasted data set without Subject 18. Click **Finish.**

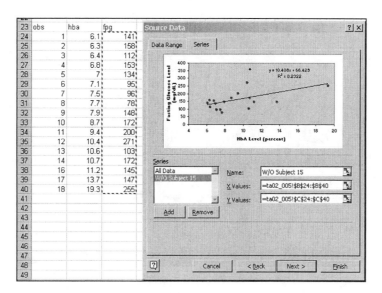

25. The trendline for the new data set stops short of the full x axis length because Subject 18 was removed. In order to observe the differences in the trendlines, we want to extend that line by right-clicking on it and selecting **Format Trendline.** Click on the **Options** tab and select **Forecast Units** of 6. This will extend the trendline to the end of the existing x axis.

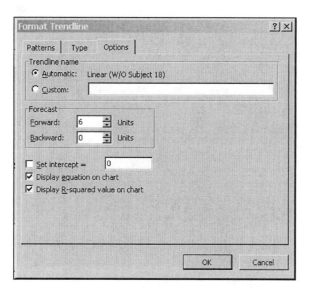

26. To avoid confusion, you can click inside the equation boxes and change the labels.

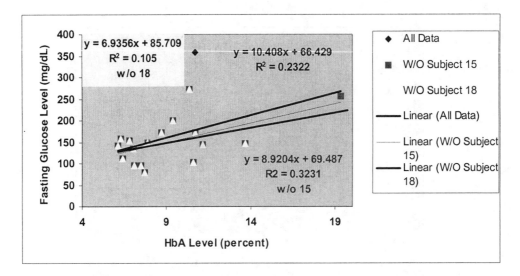

Remarks

Removing Subject 18 results in a more drastic change of the regression line, although the R^2 value decreased. This outlier gives validity to the original trendline and removing it results in increased scatter. The residual plot would not identify this influential outlier.

Chapter 3

Producing Data

Chapters 1 and 2 defined methods of data analysis using both numerical summaries and graphical techniques. Chapter 3 explores the basic methods used to produce these data sets. The objective in designing either an observational study or an experiment is to gather information about a population by gathering data from a smaller sample of that population. It is imperative that the sample be a representative snapshot of the population. Therefore, techniques of random sampling are used to ensure that no bias is introduced into the study.

A simple random sample (SRS) of size n consists of n individuals from the population chosen in such a way that every subset of size n individuals has an equal chance of being selected. The textbook demonstrates the manual method of selecting a sample by using Table B, the Random Digits Table. This chapter demonstrates computer methods to perform the same task. More complex methods are not described in this section.

3.1 First Steps

There are no specific analysis tools in Excel to address this section.

3.2 Design of Experiments

The **Sampling and Random Number Generation** tools in Excel generate samples from a specified population. Two of these tools are demonstrated using a textbook example. These tools are part of Excel's **Analysis ToolPak**. To check if the **Analysis ToolPak** is installed, select **Tools** ⇨ **Add-Ins** from the menu and then verify that the **Analysis Toolpak** is selected. If it is not selected, click on that menu option to select it. When you select **Tools** from the menu again, the option **Add-Ins** should be available.

The Sampling Analysis Tool

The **Sampling Analysis Tool** available in Excel creates a sample from a population by treating the input range as a population.

IPS Example 3.6 How to Choose an SRS—Method 1

Does talking on a hands-free phone distract drivers? Undergraduate students "drove" in a high-fidelity driving simulator equipped with a hands-free cell phone. The car ahead brakes: How quickly does the subject respond? Twenty students (the control group) simply drove. Another 20 (the experimental group) talked on the cell phone while driving. This experiment has a single factor (cell phone use) with two levels. The researchers must divide the 40 student subjects into two groups of 20. To do this in a completely unbiased fashion, put the names of the 40 students in a hat, mix them up, and draw 20. These students form the experimental group and the remaining 20 make up the control group.

1. A numerical label is assigned to each of the 40 students, starting with "01" and going up to "40". The label should have as few digits as possible but account for the highest value. In this case, there are 40 clients; therefore, two digits are appropriate, starting either with "00" or "01". For consistency with the text, 01 is used for the first client.

2. Enter the students' numbers in column A of an Excel worksheet. The label 01 can be entered as "1". Sequential numbers are generated by highlighting the first two numbers, moving the mouse pointer to the lower right of the highlighted area until you see a cross hair and then dragging down until the final number "40" is entered.

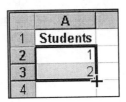

3. Select **Tools** ⇨ **Data Analysis** from the menu. Select **Sampling** and click **OK**.

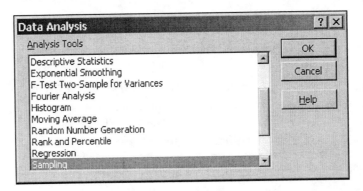

4. Select the assigned client numbers as the **Input Range**, input "20" as the **Number of Samples,** and select an empty cell to the right of the data for the **Output Range**. Click **OK**.

5. The random selection will be displayed as output. This method yields a different result each time because it is a random sample. Further details on other sampling options can be found in Excel **Help**

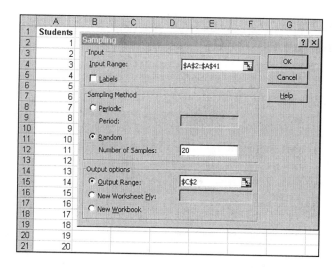

6. The students chosen for the experimental group are listed in the output. The remaining 20 students will be placed in the control group. Each time this sampling tool is run, a different sample is chosen because it is based on random sampling.

	A	B	C
1	Students		Exp Group
2	1		11
3	2		23
4	3		13
5	4		30
6	5		2
7	6		6
8	7		22
9	8		22
10	9		3
11	10		1
12	11		16
13	12		4
14	13		14
15	14		3
16	15		37
17	16		40
18	17		36
19	18		25
20	19		15
21	20		5

Remarks

Periodic sampling methods and other more complex sampling methods are not addressed in this manual.

The RAND ( ) Function

The **RAND ( )** function returns an evenly distributed random number greater than or equal to 0 and less than 1. These numbers can be sorted and a random sample chosen from the generated values.

IPS Example 3.6 How to Choose an SRS—Method 2

1. Assign a numerical label to each student as shown in **Step 1** in the previous section, **Sampling Analysis Tool.**

2. Enter the clients' numbers in column A of an Excel worksheet. The label "01" can be simply entered as "1". Sequential numbers are generated by highlighting the first two numbers, moving the mouse pointer to the lower right of the highlighted area until you see a cross hair and then dragging down until the final number "40" is entered.

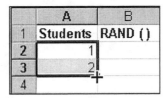

3. Enter the function "= RAND ( )" in Cell B2 and copy down to cell B41.

4. Select cells B2:B41 (the RAND values). Select **Edit** ⇨ **Copy** from the menu. The random numbers are generated. The actual random number values will change every time this exercise is performed because they are chosen at random.

5. With cells B2:B41 still selected, select **Edit** ⇨ **Paste Special** from the menu. Select **Values** and **None** and click **OK.** This process replaces the cells in column B with actual values rather than with a function. This method prevents the random numbers from changing for this example.

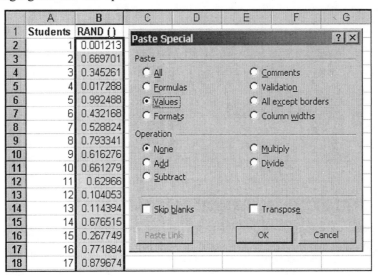

6. Select cells A2:B41 and then select **Data** ⇨ **Sort.** Under the drop down list for **Sort by**, select **RAND( ).** Verify the **Header row** is selected, and click **OK.**

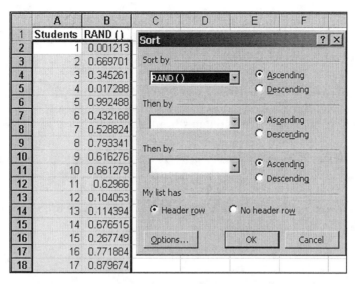

7. The random sort performed selected the sample. The first 20 numbers will be chosen as the random sample of 20 students. Remember, your students will be different because each sort produces a different random sample.

	A	B
1	Students	RAND ()
2	1	0.001213
3	23	0.002816
4	4	0.017288
5	12	0.104053
6	13	0.114394
7	20	0.128348
8	35	0.220062
9	33	0.23589
10	15	0.267749
11	31	0.294016
12	29	0.326361
13	3	0.345261
14	40	0.371122
15	30	0.407604
16	6	0.432168
17	36	0.437897
18	28	0.469007
19	18	0.485326
20	7	0.528824

RANDBETWEEN(bottom,top)

The **RANDBETWEEN** function returns a random number between two numbers that you define. A random number is returned every time the worksheet is recalculated.

IPS Example 3.6 How to Choose an SRS—Method 3

1. Assign a numerical label for each student as defined in the previous section using the Excel **Sampling Analysis Tool**.

2. Enter the function "=RANDBETWEEN(1,40)" into an empty cell in an Excel worksheet. Select the **F9** function key on your keyboard and press **Enter.** Using the F9 function key changes the cell contents from a function to value.

3. A random number has been chosen between 1 and 40. The required random sample for this problem is 20. This process will have to be repeated 20 times.

4. An alternative method is to use the **Copy** and **Paste Special** method defined earlier in the **=RAND( )** section.

3.3 Sampling Design

There are several types of sampling designs that can be used for experiments or studies. The previous section demonstrated the SRS or simple random sampling method. In this method, a table of random digits or a computer simulation is used to simulate sampling from a population, the entire group of individuals about whom we want to gather information. The SRS is chosen such that every individual has an equal chance to be chosen.

Other more complicated methods are not demonstrated in this manual.

3.4 Toward Statistical Inference

Statistical inference is the process of inferring something about a population based on a sample data set. A **parameter** is a measure of some aspect of the population—the wider group that is of interest. A **statistic** is a measure of some aspect of the sample—the smaller group chosen at random from the population. A table of random digits or a computer simulation can be used to simulate sampling from a population.

Results vary from sample to sample. This is called **sampling variability**. If many samples are taken, a particular sample statistic can be calculated, such as a proportion for each sample. The resulting proportions can be displayed graphically using a histogram. The result is a **sampling distribution**.

IPS Example 3.25 Sampling Simulation

We will simulate drawing simple random samples (SRSs) of size 100 from the population of all adult U.S. residents. Suppose that in fact 60% of the population find clothes shopping time consuming and frustrating. Then, the true value of the parameter we want to estimate is $p = 0.6$. Each digit in the random table that is generated will represent one person in this population. Digits 0 to 5 stand for people who find shopping frustrating, and 6 to 9 stand for people who do not.

1. Input the label "Set 1" into cell A1, "Set 2" into B1, and then highlight A1:B1 and drag to the right until you reach Set 50.

2. Input the function "=INT(10*RAND( ))" into cell A2, press **Enter,** and then copy the formula down to cell A101 (100 cells).

3. Highlight A2: A101 and copy the formula to the right until you reach Set 50.

4. Enter the function "=COUNTIF(A2:A101,"< 7")" into cell A103, press **Enter,** and copy to the right until you reach Set 50. This function counts the values from 0 to 6, representing the proportion of shoppers finding shopping frustrating.

5. With this row of data (A2:A101) still highlighted, select **Edit** ⇨ **Copy,** and then with cells A2:A101 still selected, select **Edit** ⇨ **Paste Special** from the menu. Select **Values** and **None,** and click **OK.** This process replaces the cells in column B with actual values rather than the function.

6. A histogram can be created from the row of proportions. It may be desired to transpose the data set to a column (select **Edit** ⇨ **Copy,** then **Edit** ⇨ **Paste Special** and then **Transpose**).

Remarks

Each time this simulation is created, the result will change due to sampling variability.

Chapter 4

Probability: The Study

of Randomness

Probability forms the basis of statistics. The laws of probability answer the questions about what happens when the same process is repeated many, many times. Inference is possible when chance is used to choose participants and treatments in a study. This chapter investigates the principles of probability used in statistical theory.

4.1 Randomness

A **statistic** is a sample value used to estimate a population **parameter**. A sample statistic such as $\bar{x}$ would be a different value every time a sample is taken. If many of these samples were taken, the mean of the sampling distribution (distribution of $\bar{x}$'s from all samples) would be the population parameter, in this case the population mean μ. In the short term, chance behavior is unpredictable but is predictable in the long run.

Random Number Generation

IPS Example 4.1 Coin Tossing

When you toss a coin, there are only two possible outcomes, heads or tails. The proportion of tosses that produce heads is quite variable at first. As we make more and more tosses, however, the proportion of heads for both trials gets close to 0.5 and stays at that value.

We are going to simulate the coin tossing by using an Excel function that yields a random number each time it is calculated.

1. Open an empty worksheet in Excel and type the following in cell A1: "Simulation of 100 Tosses of a Fair Coin".

2. Input the following equation into cell A3: "=INT(2*RAND())". The **RAND** function produces a number uniformly distributed on the (0,1) interval. The result is multiplied by 2, and then only the integer part of the number is shown. This function

will result in values 0 or 1 with equal probability. Copy the formula down to cell A102.

3. Select cells A3 to A102 and select **Edit** ⇨ **Copy** or click on the **Copy** button in the toolbar. Select **Edit** ⇨ **Paste Special** and then select **Values.** Click **OK.** This procedure copies and pastes only the values onto the selected cells without the formula. This prevents the random values from updating every time you update another item on the worksheet.

4. Input "0" into cell B2.

5. Input the equation "=A3 + B2" into cell B3. Copy the formula down to cell B102.

6. Input "0" into cell C2 and input "1" into cell C3. Fill the series down to 100 by selecting the first two numbers and dragging by the lower right (at the cross hair) until you reach 100.

7. Input "0.5" into D2 and copy down to cell D102.

8. Input the equation "= B3/C3" into cell E3 and copy down to cell E102.

9. Select the range C3 to E102 and then click on the **ChartWizard** icon in the toolbar.

10. Select the **XY Scatterplot** with data points connecting lines. Click **Next** until **Step 3.** Input the *x* axis label as "Number of Tosses" and the *y* axis label as "Proportion of Heads". Click **Finish.**

11. The resulting graph displays the probability as the number of tosses increase to 100. Re-scale the *x* axis to have a maximum value of 100. The graph below is generated for one set of probabilities. Your graph will look slightly different because it changes every time you repeat the process with new random variables. You could also increase the number of tosses and repeat the process.

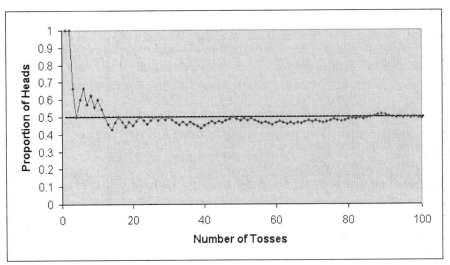

Remarks

As the number of tosses increases, the probability approaches 0.5 or 50%. The resulting probability of any outcome of a random phenomenon is the proportion of times the outcomes would occur in a very long series of repetitions.

4.2 Probability Models

A **probability model** includes a mathematical description of a random phenomenon consisting of a sample space S of all possible outcomes and an outcome (or a set of outcomes) of a random phenomenon called an event. The event represents a subset of the sample space. The sample space can be as simple as heads or tails for a coin toss or as complex as 10,000 randomly chosen households to be surveyed out of a sample space of 100 million.

Creating a Probability Histogram

IPS Example 4.10 Random Digits

You might think that first digits are distributed "at random" among the digits 1 to 9. The nine possible outcomes would then be equally likely. The sample space for a single digit is:

$$S = \{1, 2, 3, 4, 5, 6, 7, 8, 9\}$$

Call a randomly chosen first digit X for short. The probability model for X is completely described by this table:

First digit X	1	2	3	4	5	6	7	8	9
Probability	1/9	1/9	1/9	1/9	1/9	1/9	1/9	1/9	1/9

1. Open an empty worksheet in Excel and type the following in cell A1: "Random Digits".

2. Input the following equation into cell A2: "=1+ INT(9*RAND())". The RAND function produces a number uniformly distributed on the (1, 9) interval. The result is multiplied by 9 and only the integer part of the number is shown. This will result in values of 1 to 9 with equal probability. Copy the formula down to cell A101.

3. Select cells A2 to A101 and select **Edit** ⇨ **Copy** from the menu or click the **Copy** button in the toolbar. Select **Edit** ⇨ **Paste Special** from the menu and then select **Values.** Click **OK.** This procedure copies and pastes the same values into the cells without the formula. This procedure prevents the random values from updating every time you update another item on the worksheet.

4. Input the numbers 1 to 9 starting in cells C1 to C9 as the histogram bin sizes.

5. Select **Tools** ⇨ **Data Analysis** from the menu. Select **Histogram** and click **OK.** Select the random numbers as the **Input Range** and the numbers 1 through 9 as the **Bin Range.** Select **Output Range** and click into cell C12. Do **NOT** select **Chart Output.** Click **OK.**

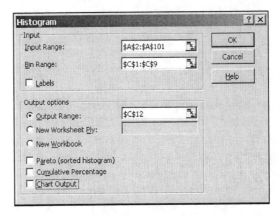

6. Input the formula "=D13/100" into the cell E13. Copy the formula down to cell E21. This formula produces the proportion out of 100 for each number.

7. Select the resulting proportions (only the *y* variables need to be selected because the *x* variables are numbered in order starting with one, which is the default). Select the **ChartWizard** icon in the toolbar. Select **Next** until **Step 3** and input the *x* axis title as **Outcomes** and the *y* axis label as **Probability.**

8. Select the **Legend** tab and de-select the Legend. Click **Finish.**

9. Right-click on any bar in the graph, select **Format Data Series,** select the **Options** tab, change the **Gap Width** to "0", and click **OK.**

10. The resulting bar graph should look similar to the one below. However, each time the random numbers are created, the resulting sample will have a different probability distribution.

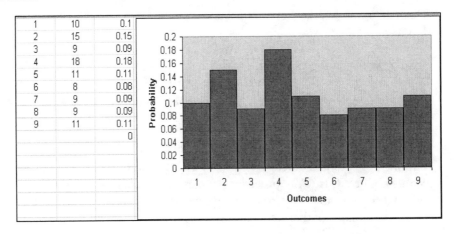

Remarks

For a finite number of outcomes, a probability is assigned to each individual outcome. In this example, there is an equal chance that each number is chosen; therefore, the probability is the same for each number—1/9. In our simulation of this example, there was variability around the probability of 0.11. If the sample size increases to 1000, the probability for each number would again approach 0.11. The following probability histogram was created with a sample size of 1000.

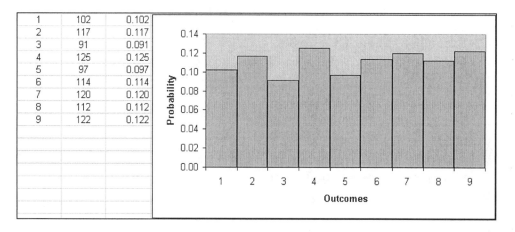

1	102	0.102
2	117	0.117
3	91	0.091
4	125	0.125
5	97	0.097
6	114	0.114
7	120	0.120
8	112	0.112
9	122	0.122

4.3   Random Variables

A sample space doesn't always have to contain numbers. For example, in our coin-tossing example, the sample space can be defined by H for heads and T for tails rather than 0 or 1. If X is the number of heads, tossing a coin multiple times will yield X values of 0, 1, 2, ... up to the maximum number of tosses. X is called a **random variable** because its value varies when the coin toss is repeated. A **random variable** is defined as a variable whose value is a numerical outcome of a random phenomenon, in our example, coin tossing.

In our previous example of random number generation for numbers between 1 and 9, the possible values for X are the nine number values. In this case, X is a discrete random variable. This definition is opposed to a sample space that could contain any values from 0 to 1 and therefore, defined as a continuous random variable Y.

The **probability distribution** of a random variable X defines the values for X and the probabilities for those variables.

Discrete Random Variables

A **discrete random variable** X has a finite number of possible values. The probability distribution of X lists all possible values of X and their probabilities of occurrence. Every individual probability is between 0 and 1, and the sum of the probabilities is equal to 1. A **probability histogram** is a graphical representation of the probabilities of all possible outcomes.

IPS Example 4.17 Four Coin Tosses

What is the probability distribution of the discrete random variable X that counts the number of heads in four tosses of a coin? Toss a balanced coin four times. There are 16 possible outcomes summarized by:

$X = 0$ TTTT
$X = 1$ HTTT, THTT, TTHT, TTTH
$X = 2$ HTTH, HTHT, THTH, HHTT, THHT, TTHH
$X = 3$ HHHT, HHTH, HTHH, THHH
$X = 4$ HHHH

The resulting probabilities are defined by:
$P(X = 0) = 1/16 = 0.0625$
$P(X = 1) = 4/16 = 0.25$
$P(X = 2) = 6/16 = 0.375$
$P(X = 3) = 4/16 = 0.25$
$P(X = 4) = 1/16 = 0.0625$

The probability histogram for this distribution is created by graphing the results in Excel. In this case, the distribution represents the symmetric probability after many tosses of the four coins.

# Heads	Probability
0	0.063
1	0.250
2	0.375
3	0.250
4	0.063

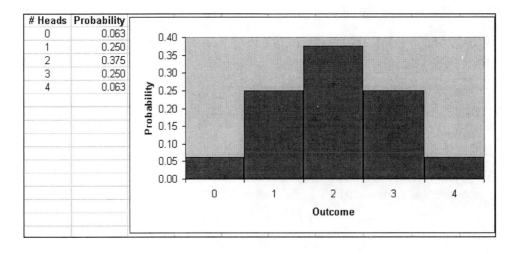

Remarks

The probability histogram reflects the probability of getting heads in four tosses for many samples. For each outcome, the probabilities are added. For example, for $X = 1$, the probabilities are added for each outcome with only one head in the toss—HTTT, THTT, TTHT, TTTH. Exactly one head occurs four times out of 16 possibilities. Therefore, the probability $P(X = 1) = 4/16 = 0.25$.

Continuous Random Variables

A **continuous random variable** X takes all values in an interval of numbers. The probability distribution of X is described by a density curve. The probability of any event is the area under the density curve and above the values of X that make up the event. The results of many trials are represented by the density curve of a **uniform distribution.** The probably of any event is the area under the density curve and above the event in question.

Normal Distributions as Probability Distributions

The density curves that are most familiar to us are the normal curves. Density curves assign probabilities and, therefore; normal distributions are probability distributions. If X has a normal distribution with mean μ and stardard deviation σ, denoted by $N(\mu, \sigma)$, then we can calculate the standardized variable using the following equation:

$$z = \frac{X - \mu}{\sigma}$$

Where z is a standard normal random variable having the distribution $N(\mu, \sigma)$.

IPS Example 4.19 Reporting Cheating

Students are reluctant to report cheating by other students. A sample survey puts this question to an SRS of 400 undergraduates: "You witness two students cheating on a quiz. Do you go to the professor?" Suppose that if we could ask all undergraduates, 12% would answer "Yes."

The proportion $p = 0.12$ is a parameter that describes the population of all undergraduates. The proportion $\hat{p}$ of the sample who answer "Yes" is a statistic used to estimate p. The mean 0.12 of the distribution is the same as the population parameter. The standard deviation is controlled mainly by the size of the sample and can be assumed to be 0.016.

What is the probability that the survey result differs from the truth about the population by more than two percentage points? Because $p = 0.12$, the survey would result in $\hat{p} < 0.10$ or $\hat{p} > 0.14$.

We will use the solution method previously used in Chapter 1.

1. Type in the given values and their headings into an empty worksheet. We are interested in the proportion or probability of having a value as great as or greater than 0.14. However, because of how the normal distribution table (Table A in the text) is constructed, the area to the left of a specific value is always calculated. If an area to the right of a specific value is desired, we will have to subtract the value from 1. In addition, we are interested in the probability of having a value as small or smaller than 0.10.

Pop Mean	0.12
Pop SDev	0.016
x1	0.10
x2	0.14
z-score	
P < 0.14	
P > 0.14	
z-score	
P < 0.10	
Total	

2. Click in the cell to the right of the **z-score** to input the function to calculate the standardized z-value for a specific x-value.
    Input "= STANDARDIZE(0.14,0.12,0.016)" and press **Enter.** The arguments of the function in order are the x-value, the population mean, and the population standard deviation. Instead of typing in the specific values, you can click on the cells containing those values. The result is 1.25.

3. Click in the cell to the right of the **P < 0.14** cell to input the function to calculate the probability of an x-value being as low or lower than 0.14.
    Input "= NORMSDIST(1.25)" and press **Enter.**

4. Click in the cell to the right of the **P > 0.14** cell to input the function to calculate the probability of an x-value being as high or higher than 0.14.
    Input "= 1 – ", click on the cell above (0.89435) and press Enter. This calculates the area to the right of 0.14 as 0.10565.

Pop Mean	0.12
Pop SDev	0.016
x1	0.10
x2	0.14
z-score	1.25
P < 0.14	0.89435
P > 0.14	0.10565

5. Click in the cell to the right of the second **z-score** cell to input the function to calculate the standardized *z*-value for < 0.10.
   Input "= STANDARDIZE(0.10,0.12,0.016)" and press **Enter.** The result is –1.25.

6. Click in the cell to the right of the **P < 0.10** cell to input the function to calculate the probability of an *x*-value being as low or lower than 0.10.
   Input "= NORMSDIST(–1.25)" and press **Enter.** The result is 0.10565.

z-score	1.25
P < 0.14	0.89435
P > 0.14	0.10565
z-score	-1.25
P < 0.10	0.10565

7. Add the probabilities to get the total for **P < 0.10** and **P > 0.14.** The result is 0.2113.

Remarks

The resulting proportion equals 0.2113, or approximately 21% will be off by more than 2 percentage points. That is the proportion of students that would yield a statistic of $\hat{p} < 0.10$ or $\hat{p} > 0.14$.

4.4 Means and Variances of Random Variables

The means of random variables behave like averages. For a single random variable *X*, the linear regression line that describes these averages consists of an intercept *a* and slope *b* of the form:

$$\mu_{a+bX} = a + b\mu_X$$

If *X* and *Y* are both random variables, the regression equation becomes:

$$\mu_{X+Y} = \mu_X + \mu_Y$$

The Mean of a Random Variable

Probability describes the behavior of a random phenomenon over the long run. The probability distribution of a random variable reflects the distribution of many observations. The distribution looks similar to the normal density curves previously studied. Next, we will expand our analysis to include the mean of all possible values of *X*.

We define X as a discrete random variable with distribution defined as:

Value of X	x_1	x_2	x_3	...	x_k
Probability	p_1	p_2	p_3	...	p_k

The mean of a probability distribution is μ, the same as the mean of a population distribution because, over the long run, the probability distribution will approach the population distribution.

To find the mean of X, each possible value of X is multiplied by its probability. The products are then all added:

$$\mu_X = x_1 p_1 + x_2 p_2 + \cdots + x_k p_k = \sum x_i p_i$$

IPS Example 4.24 Car Dealership Sales

Linda is a sales associate at a large auto dealership. At her commission rate of 25% of gross profit on each vehicle she sells, Linda expects to earn $350 for each car sold and $400 for each truck or SUV sold. Linda motivates herself by using probability estimates of her sales. For a sunny Saturday in April, she estimates her car sales as follows:

Cars sold	0	1	2	3
Probability	0.3	0.4	0.2	0.1

Linda's estimate of her truck or SUV sales is:

Vehicles sold	0	1	2
Probability	0.4	0.5	0.1

Let X be the number of cars Linda sells and Y the number of trucks or SUVs. The means of these random variables are:

$$\mu_X = (0)(0.3) + (1)(0.4) + (2)(0.2) + (3)(0.1) = 1.1 \text{ cars}$$

$$\mu_Y = (0)(0.4) + (1)(0.5) + (2)(0.1) = 0.7 \text{ trucks or SUVs}$$

Linda's earnings, at $350 per car and $400 per truck or SUV, are:

$$Z = 350X + 400Y$$

Her mean earnings are:

$$\mu_Z = (350)\mu_X + (400)\mu_Y = (350)(1.1) + (400)(0.7) = \$665$$

Statistical Estimation and the Law of Large Numbers

Sample statistics such as a mean $\bar{x}$ are random variables. This statistic can be used to estimate the population parameter μ. Every individual sample mean will be different, however, as the number of samples increases, the mean $\bar{x}$ should be very close to the population mean μ.

If independent observations are taken at random from any population with mean μ, as the number of observations increases, the mean $\bar{x}$ approaches the population mean μ. This is referred to as the **Law of Large Numbers.**

We can illustrate this law by revisiting Example 4.1 and solving the problem for the probability of obtaining heads in a coin toss. In the previous example, we investigated increased sample sizes by calculating the accumulated proportion of heads as we increase the sample size. The result was that the accumulated proportion approached 0.5 over the long run.

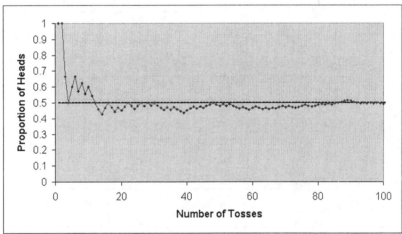

We will now investigate the same problem by using a different method. Instead of increasing the sample size and calculating the accumulated proportion, we will create 50 samples of 100 tosses each and find the mean $\bar{x}$ of each sample. A sampling distribution will be created from these mean values. The center of this distribution should be the center of the population.

The Variance of a Random Variable

The mean is the center of a distribution. A complete description of that distribution requires knowledge of the spread or variability of the distribution. The variance of a random variable X is written as σ^2_X. The definition of the variance σ^2_X of a random variable is similar to the definition of the sample variance s^2 given in Chapter 1 of the textbook. The variance is an average value of the squared deviation $(X - \mu_X)^2$ of the variable X from its mean μ_X. The standard deviation of σ_X of X is the square root of the variance.

IPS Example 4.25 Car Dealership Sales Continued

In Example 4.24, we saw that the number X of cars that Linda hopes to sell has the following distribution:

Cars sold	0	1	2	3
Probability	0.3	0.4	0.2	0.1

We can find the mean and variance of X by arranging the calculation in a table that we can input into an Excel worksheet.

1. Input the labels and given values as shown below. x_i represents the number X of cars and p_i is the probability of each value.

	A	B	C	D
1	x_i	p_i	$x_i p_i$	$(x_i - u_X)^2 p_i$
2	0	0.3		
3	1	0.4		
4	2	0.2		
5	3	0.1		

2. Input the formula in cell C1, as shown below, to multiply x_i by p_i. Copy the formula down to cell C5. Add the values in column C to calculate μ_X.

	A	B	C	D
1	x_i	p_i	$x_i p_i$	$(x_i - u_X)^2 p_i$
2	0	0.3	=A2*B2	
3	1	0.4	=A3*B3	
4	2	0.2	=A4*B4	
5	3	0.1	=A5*B5	
6		u_X	=SUM(C2:C5)	

3. Input the formula in cell D2, as shown below to calculate the variance. Copy the formula down to cell D5. Add the values in column D to calculate σ^2_X.

	A	B	C	D
1	x_i	p_i	$x_i p_i$	$(x_i - u_X)^2 p_i$
2	0	0.3	=A2*B2	=(A2-1.1)^2*B2
3	1	0.4	=A3*B3	=(A3-1.1)^2*B3
4	2	0.2	=A4*B4	=(A4-1.1)^2*B4
5	3	0.1	=A5*B5	=(A5-1.1)^2*B5
6		u_X	=SUM(C2:C5)	=SUM(D2:D5)

4. The values calculated are shown below:

	A	B	C	D
1	x_i	p_i	$x_i p_i$	$(x_i - u_X)^2 p_i$
2	0	0.3	0.0	0.363
3	1	0.4	0.4	0.004
4	2	0.2	0.4	0.162
5	3	0.1	0.3	0.361
6		u_X	1.1	0.89

5. The standard deviation of X is the square root of the calculated sum of 0.89. Input the **SQUARE ROOT** function as shown below to calculate the standard deviation of 0.943.

	A	B	C	D
1	x_i	p_i	$x_i p_i$	$(x_i - u_X)^2 p_i$
2	0	0.3	=A2*B2	=(A2-1.1)^2*B2
3	1	0.4	=A3*B3	=(A3-1.1)^2*B3
4	2	0.2	=A4*B4	=(A4-1.1)^2*B4
5	3	0.1	=A5*B5	=(A5-1.1)^2*B5
6		u_X	=SUM(C2:C5)	=SUM(D2:D5)
7			StDev	=SQRT(D6)

6. The values calculated are shown below:

	A	B	C	D
1	x_i	p_i	$x_i p_i$	$(x_i - u_X)^2 p_i$
2	0	0.3	0.0	0.363
3	1	0.4	0.4	0.004
4	2	0.2	0.4	0.162
5	3	0.1	0.3	0.361
6		u_X	1.1	0.89
7			StDev	0.943

Remarks

The standard deviation is a measure of the variability of the number of cars Linda sells.

Chapter 5

Sampling Distributions

The mathematical reasoning behind probability is discussed in this chapter, where we will provide the necessary background and understanding of basic probability rules with applications to the principles of statistics.

5.1   Sampling Distributions for Counts and Proportions

The general rules of probability are defined in the textbook. Once understood, more complex random phenomena can be modeled. One important underlying principle for the multiplication rule is independence, that is, the outcome of one event does not influence the outcome of another event. The probability of drawing any card in a deck is 1/52. However, if a card is drawn and kept out of the deck, the next drawing would not be independent of the first.

The types of problems outlined in this section of the text can be done manually and do not require the use of Excel.

The Binomial Distributions

A binomial distribution is a probability model for a *count* of successful outcomes. In this model, *n,* the numbers of observations are all independent and each observation is a discrete response, corresponding to "success" or "failure." The probability of success *p* is the same for each observation.

If random variable X (number of successes) has a binomial distribution, the possible values of X are 0, 1, 2, …, n. If k is any one of these values,

$$P(X = k) = \binom{n}{k} p^k (1 - p)^{n-k}$$

with mean $\mu_X = np$ and standard deviation $\sigma = \sqrt{np(1 - p)}$.

Creating a Binomial Distribution

IPS Example 5.3 Financial Records Audit

The financial records of businesses may be audited by state tax authorities to test compliance with tax laws. It is too time-consuming to examine, for example, all sales and purchases made by a company during the period covered by the audit. The auditor will examine samples chosen by judgment though they are often chosen at random. Suppose that the audit chose just 15 sales records. What is the probability that no more than 1 of 15 is misclassified? The count X of misclassified records in the sample has approximately the B (15, 0.08) distribution. The binomial distribution for this example can be modeled in Excel with $n = 15$ and $p = 0.08$.

1. Open an empty worksheet in Excel and type the following into cell A1: "Binomial Distribution".

2. Input the label *"k"* in cell A3, the label "P($X = k$)" into cell B3 and the label "P($X <= k$)" in cell C3.

3. Input "0" into cell A4, "1" into cell A5, and fill down to 15 in cell A19.

4. Input the following function into cell B4: "= BINOMDIST(A4, 15, 0.08, 0)" and press **Enter.** This function calculates the exact binomial probability that *k* number of records are found misclassified based on the total number in the sample (15), the given population parameter of 8% misclassified. The last argument (set to 0) specifies whether to calculate the exact probability for that record count or a cumulative probability. Copy the function down to cell B19.

5. Input the following function into cell C4: "= BINOMDIST(A4, 15, 0.08, 1)" and press **Enter.** This function is identical to the previous function with the exception of the last argument (now set to 1), which calculates the cumulative probability. Copy the function down to cell C19.

6. The resulting table of individual and cumulative probabilities is shown below:

	A	B	C
1	Binomial Distribution		
2			
3	k	P (X=k)	P (X<=k)
4	0	0.286297	0.286297
5	1	0.373431	0.659729
6	2	0.227306	0.887035
7	3	0.085652	0.972686
8	4	0.022344	0.99503
9	5	0.004274	0.999305
10	6	0.000619	0.999924
11	7	6.93E-05	0.999994
12	8	6.02E-06	1
13	9	4.07E-07	1
14	10	2.13E-08	1

7. The values in cells B11 to B19 could display scientific notation for a value that is almost zero (10^{-6}, 10^{-7}, 10^{-9}). You can keep the values as they appear or re-format them to fewer decimals by first selecting those cells, then selecting **Format** ⇨ **Cells** from the menu. On the **Number tab**, select **Number** for the format and input "4" for the number of **Decimal Places.**

8. The resulting data table is in the correct format to create a bar graph that would be the same as a histogram in this case. Select cells B4:B19 and click the **ChartWizard** in the toolbar. Select **Column** chart and click **Next.** Click the **Series** tab. Click in the box to the right of **Category (X) axis labels** and select the *k* values in cells A4:A19. Click **Next.**

9. Input "Count of Misclassified Records" for the **Category (X) axis label** and "Probability" for the **Value (Y) Axis label.** Click the **Legend** tab and de-select **Show Legend.** Click **Finish.**

10. Right-click on the gridlines and select **Clear.**

11. Right-click on any column and select **Format Data Series,** the **Options** tab, and change the **Gap Width** to "0". Click **OK.** The resulting histogram is shown below:

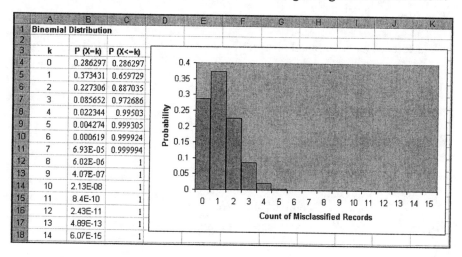

The Normal Approximation to Binomial Distribution

The binomial distribution methods are only practical for a small number of trials. As the number of trials *n* gets larger, the binomial distribution approaches the normal distribution.

IPS Example 5.7 The Helsinki Heart Study

The Helsinki Heart Study asked whether the anticholesterol drug gemfibrozil reduces heart attacks. In planning such an experiment, the researchers must be confident that the sample sizes are large enough to enable them to observe enough heart attacks. The Helsinki study planned to give gemfribrozil to about 2000 men aged 40 to 55 and a placebo to another 2000. The probability of a heart attack during the five-year period of the study for men is about 0.04. What are the mean and standard deviation of the number of heart attacks that will be observed in one group if the treatment does not change this probability?

The responses can be assumed to be independent. The random variable X has a binomial distribution with $n = 2000$ and $p = 0.04$. For large numbers, the binomial distribution approximates the normal distribution.

First, the mean μ and the standard deviation σ are calculated by using the binomial distribution:

$$\mu = np = (2000)(0.04) = 80$$

$$\sigma_x = \sqrt{np(1-p)} = \sqrt{(2000)(0.04)(0.96)} = 8.76$$

Remarks

When count X has a binomial distribution with n trials and probability p, the distribution is approximately normal and denoted by $N(np, \sqrt{np(1-p)})$. A normal approximation can be used when n and p satisfy $np \geq 10$ and $n(1-p) \geq 10$.

Sample Proportions

An estimate of a proportion p of "successes" in a population is given by:

$$\hat{p} = \frac{\text{count of successes in sample}}{\text{size of sample}} = \frac{X}{n}$$

The mean and standard deviation of $\hat{p}$ are defined as $\mu_{\hat{p}} = p$ and $\sigma_{\hat{p}} = \sqrt{\dfrac{p(1-p)}{n}}$. The formula for the standard deviation is approximately correct for an SRS from a large population of at least 20 times as large as the sample.

Normal Approximation for Counts and Proportions

When n is large, the sampling distribution for counts and proportions are approximately normal and defined by the following:

X is approximately $N(np, \sqrt{np(1-p)})$ and $\hat{p}$ is approximately $N\left(p, \sqrt{\dfrac{p(1-p)}{n}}\right)$ for values of n and p that satisfy $np \geq 10$ and $n(1-p) \geq 10$.

IPS Example 5.11 Financial Records Audit Continued

The audit described in Example 5.3 examined an SRS of 150 sales records for compliance with sales tax laws. In fact, 8% of all the company's sales records have an incorrect sales tax classification. With the sample size at 150, the shape of the binomial distribution is approximately normal.

$$\mu_X = np = (150)(0.08) = 12$$

$$\sigma_X = \sqrt{np(1-p)} = \sqrt{(150)(0.08)(0.92)} = 3.3226$$

Calculate the normal approximation for the probability of no more than 10 misclassified records.

1. Input the following labels and given values into Excel:

	A	B
1	p	0.08
2	n	150
3	u_x	
4	σ_x	
5		
6	z-score	
7	P ≤ 10	

2. Input the following formulas to calculate the mean and standard deviation of the count X:

	A	B
1	p	0.08
2	n	150
3	u_x	=B1*B2
4	σ_x	=SQRT(B3*(1-B1))

3. The resulting values are shown below:

	A	B
1	p	0.08
2	n	150
3	u_x	12
4	σ_x	3.3226

4. Input the z-score and probability formulas as shown below:

	A	B
6	z-score	=((10-B3)/B4)
7	P ≤ 10	=NORMSDIST(B6)

5. The resulting values are shown below:

6	z-score	-0.602
7	P ≤ 10	0.274

Remarks

The binomial probability of $P(X \leq 10) = 0.3384$. The normal approximations is only roughly accurate.

5.2   The Sampling Distribution of a Sample Mean

Sample statistics such as a mean $\bar{x}$ are random variables. This statistic can be used to estimate the population parameter μ. Every individual sample mean will be different; however, as the number of samples increases, the mean $\bar{x}$ should be very close to the population mean μ.

To illustrate the point, we will create 50 samples of 100 tosses each and find the mean $\bar{x}$ of each sample. A sampling distribution will be created from these mean values. The center of this distribution should be the center of the population.

Creating a Sampling Distribution

1. Open a new worksheet and input the following formula into cell A1: "=INT(2*RAND())".

2. Copy the formula over to cell AX1 (50 columns to the right).

3. With the range A1:AX1 selected, copy the entire range of formulas down to cell AX100 by dragging the cross hair in the lower right of the AX1 cell.

4. Input the function to calculate a mean in cell A102: "=AVERAGE(A1:A100)" and press **Enter** on the keyboard.

5. Copy the function to the right 50 columns. This is the calculation of the sample mean $\bar{x}$ for each sample of 100.

6. Create a histogram by selecting **Tools** ⇨ **Data Analysis** from the menu, selecting **Histogram,** and clicking **OK.**

7. Select all of the mean values (range A102:AX102) for the **Input Range.** Select an **Output** cell, **Chart Output,** and click **OK.**

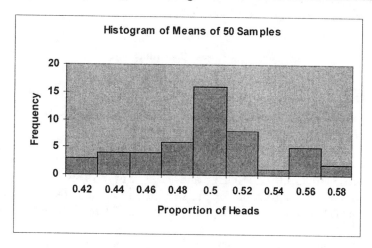

8. The resulting histogram should be enhanced by resizing, increasing the bar width (right-click on any bar and select **Format Data Series, Options** tab, and input "0" for the **Gap Width**), and formatting the bin sizes to 2 or 3 decimal places. Your histogram should look similar to the one on the following page but all samples taken were random and, therefore, the resulting means are different for each simulation.

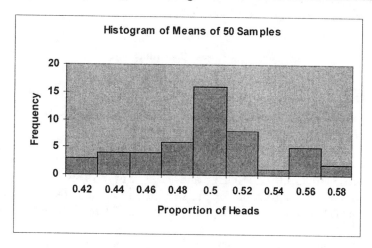

The Mean and the Standard Deviation of $\bar{x}$

If $\bar{x}$ is the mean of a random sample of size n taken from a large population with mean μ and standard deviation σ, then the mean of the sampling distribution of $\bar{x}$ is μ and the standard deviation is $\sigma/\sqrt{n}$. Further, we can say that the statistic $\bar{x}$ is an unbiased estimator of the parameter μ.

The standard deviation of the distribution of $\bar{x}$ is smaller than the standard deviation of individual observations. That is, if n is large, the standard deviation of $\bar{x}$ is small and the results are less variable.

The Central Limit Theorem

If a random sample of size n is taken from a population with mean μ and standard deviation σ, when n is large, the sampling distribution of the sample mean $\bar{x}$ is approximately normal with distribution $N\left(\mu, \dfrac{\sigma}{\sqrt{n}}\right)$.

IPS Example 5.18 Maintaining Air Conditioners

The time X that a technician requires to perform preventive maintenance on an air-conditioning unit is governed by an exponential distribution. The mean time is $\mu = 1$ hour and the standard deviation is $\sigma = 1$ hour. Your company operates 70 of these units. What is the probability that their average maintenance time exceeds 50 minutes?

The central limit theorem says that the sample mean time $\bar{x}$ (in hours) spent working on 70 units has approximately the Normal distribution with mean equal to the population mean $\mu = 1$ hour and standard deviation equal to:

$$\frac{\sigma}{\sqrt{70}} = \frac{1}{\sqrt{70}} = 0.12 \text{ hour}$$

The distribution of $\bar{x}$ is approximately $N(1, 0.12)$. Fifty minutes is 50/60 of an hour or 0.83 hour. We are interested in the probability of $\bar{x} > 0.83$. We will use the normal distribution calculations to determine the probability.

1. Open a new worksheet and input the following formula into cell A1: "=NORMDIST(0.83,1,0.12,TRUE)" and press **Enter.** This function returns the normal cumulative distribution for the specified mean and standard deviation. The value tested is 0.83 (50 min as decimal of an hour). The resulting value is 0.078.

2. Subtract this value from 1 to calculate the proportion ≥ 0.83. The resulting value is 0.922. The actual value is very close to the normal approximation.

Chapter 6

Introduction to Inference

Statistical inference allows us to draw conclusions about a population from the sample data gathered. These conclusions use probability to describe random variations in the data. The methods of formal inference are based on sampling distributions that can be defined by a probability model for the data. This model is reliable only for a properly designed random sample or randomized experiment and does not account for design flaws.

An important population parameter is the mean μ. The inference procedures used in this chapter are used to find the confidence interval and perform significance tests for the mean of a population. The procedures require use of a normal distribution and are therefore valid only when the associated population is approximately normal or the sample size is so large the distribution can be assumed normal by utilizing the central limit theorem.

6.1   Estimating with Confidence

The entire group of interest to a researcher is defined as the population. The mean of the population is defined as μ. If μ is not known, it can be estimated as a value within an interval that can be defined by several known factors:

- A sample mean $\bar{x}$ where the sample has been randomly taken from the population of interest

- The population standard deviation σ, which is not usually known but can be estimated from the sample standard deviation s for large samples

- The confidence level C that defines the probability that the interval will capture the true parameter in repeated samples

The **confidence interval** for a population mean is defined as

estimate +/– margin of error

The estimate is our guess for the value of the unknown parameter μ, which is $\bar{x}$.

In equation form, the **confidence interval** is defined as

$$\bar{x} \pm z^* \frac{\sigma}{\sqrt{n}}$$

z^* is defined as the critical z value defined by the confidence level C (the area between $-z^*$ and z^* using a normal curve equals C).

The **margin of error** m is defined as $m = z^* \dfrac{\sigma}{\sqrt{n}}$, resulting in a **confidence interval** for μ of $\bar{x} \pm m$.

The **confidence interval** shows how precise we believe our guess to be, based on the variability of the estimate. A 95% confidence interval will capture the unknown parameter μ in 95% of all possible samples.

The interval is exact when the population distribution is normal and is approximately correct when n, the sample size, is large.

IPS Example 6.3 The National Student Loan Survey

The National Student Loan Survey collects data to examine questions related to the amount of money that borrowers owe. A sample of 1,280 borrowers who began repayment of their loans at least six months prior to the study but not more than four months prior. The mean of the debt for undergraduate study was $18,900 and the standard deviation was about $49,000. This distribution is clearly skewed, but because our sample size is quite large, we can rely on the central limit theorem to assure us that the confidence interval based on the normal distribution will be a good approximation.

Let's compute a 95% confidence interval for the true mean debt for all borrowers. Although the standard deviation is estimated from the data collected, we will treat it as a known quantity for our calculation here.

The survey data are assumed to come from a normally distributed population $N(\mu, \sigma)$ with mean μ and standard deviation σ.

Calculating a Confidence Interval

The Excel function to compute a confidence interval is defined as

$$= \text{CONFIDENCE}\,(\alpha,\, \sigma,\, n)$$

where α is the confidence level, σ is the population standard deviation, and n is the sample size.

1. Input the following into an empty cell in an Excel worksheet:

=CONFIDENCE(0.05,49000,1280)

2. Press **Enter**. The result is the margin of error: 2684.346

3. The confidence interval can be calculated manually as: 18,900 +/– 2700. (the mean is rounded to the nearest $100 so the margin of error should be rounded)

4. Each component in the confidence interval can be calculated individually by subtracting the margin of error from and adding the margin of error to the sample mean. The resulting interval is (16,200, 21,600).

Remarks

The interval between 16,200 and 21,600 contains the true population mean μ, 95% of the time with repeated samples.

IPS Example 6.4 The National Student Loan Survey Continued

Let's assume that the sample mean of the debt for undergraduate study is $18,900 and the standard deviation was about $49,000 as in Example 6.3. But suppose that the sample size is only 320. The margin of error for 95% confidence is determined by the following:

1. Input the following into an empty cell in an Excel worksheet. The first argument in the function is the probability, the second argument is the standard deviation, and the third argument is the number of observations.

=CONFIDENCE(0.05,49000,320)

2. Press **Enter**. The result is the margin of error: 5368.693

3. The confidence interval can be calculated manually as: 18,900 +/– 5,400. (The mean is rounded to the nearest $100 so the margin of error should be rounded)

4. The resulting interval is (13,500, 24,300).

Remarks

The new sample size is exactly one-fourth of the original 1,280. The margin of error is doubled when the sample size is reduced to one-fourth of the original value. The interval between 13,500 and 24,300 contains the true population mean μ, 95% of the time with repeated samples.

IPS Example 6.5 The National Student Loan Survey Cont.

Suppose that for the student loan data in Example 6.3 we wanted 99% confidence, $z^* =$ 2.576. The margin of error for a 99% confidence level based on 1,280 observations can be determined by the following method:

1. The normal critical value z^* is calculated by inputting the function

$$\boxed{\texttt{=NORMSINV(0.995)}}$$

The probability of 0.995 is used instead of 0.01 because, for a 99% confidence interval, a middle area of 99% in a normal distribution leaves 0.005 on each side. The positive critical z value has an area to the left of 0.9795. The resulting critical value is 2.576.

2. Input the following function into an empty cell in an Excel worksheet. The first argument in the function is the probability, the second argument is the standard deviation and the third argument is the number of observations.

$$\boxed{\texttt{=CONFIDENCE(0.01,49000,1280)}}$$

3. Press **Enter**. The result is the margin of error. $\boxed{3527.841}$

4. The confidence interval can be calculated manually as: 18900 +/– 3500. (The mean is rounded to the nearest $100 so the margin of error should be rounded)

5. The resulting interval is (15,400, 22,400).

Remarks—How Confidence Intervals Behave

The resulting 99% confidence interval is $15,400 to $22,400. The margin of error is $3,500, which is much larger than the $2,700 margin of error calculated for the 95% confidence level.

As a confidence level increases, the margin of error and confidence interval also increases. If a higher confidence level is desired, an increased margin of error must be accepted.

As the sample size increases, the margin of error decreases and the interval also decreases.

As the standard deviation decreases, the margin of error decreases and the interval also decreases. If the standard deviation or spread of a dataset can be controlled through more restrictive measurement processes, the margin of error would decrease as well as the confidence interval.

Choosing a Sample Size

The margin of error of a confidence interval for a normal population mean is $z^* \dfrac{\sigma}{\sqrt{n}}$. If this expression is algebraically rearranged, a specific sample size can be determined for a specific margin of error.

To obtain a desired sample size for a given margin of error m, solve the equation

$$n = \left(\frac{z^* \sigma}{m} \right)^2$$

IPS Example 6.6 How Many Students Should We Survey?

Suppose that we are planning a student loan survey similar to the one described in Example 6.3. If we want the margin of error to be \$2,000 with 95% confidence, how large a sample is required?

1. The critical z-value can be determined using a previous method or by looking its value up in a table. The 95% z^* is defined as 1.96. The standard deviation is given as \$49,000 and the desired margin of error is \$2,000.

2. Input the following equation into an empty cell: $\boxed{\text{=((1.96*49000)/2000)^2}}$

3. Press **Enter**. The result is the sample size. $\boxed{2{,}305.92}$

4. If there is any decimal after the last whole digit, the number is increased by one to ensure that the desired margin of error is achieved. That increases our sample size to 2,306.

Beyond the Basics: The Bootstrap

The **bootstrap** is a new procedure for approximating sampling distributions when theory cannot tell us their shape. The basic idea is to act as if the sample were the population. We take many samples from the given sample. Each one of these is called a **resample.** The mean $\bar{x}$ is calculated for each resample. The results from the different resamples will be different because we sample with replacement. An individual observation in the original sample can appear more than once in the resample.

Simulate a Bootstrap Distribution

To demonstrate the resampling method, we will use the small sample from the text.

| 190.5 | 189.0 | 195.5 | 187.0 |

1. Input the data into an empty worksheet into a column. In the next column, calculate the individual sampling probability for each number (1/4).

	A	B
1	**Bootstrap**	**Simulation**
2	190.5	0.25
3	189.0	0.25
4	195.5	0.25
5	187.0	0.25

2. Select **Tools** ⇨ **Data Analysis** ⇨ **Random Number Generation** from the menu and input the **Number of Variables** as 4, the **Number of Random Numbers** as 20, and the **Distribution** as Discrete. Select the **Input Range** as shown and select an **Output** cell.

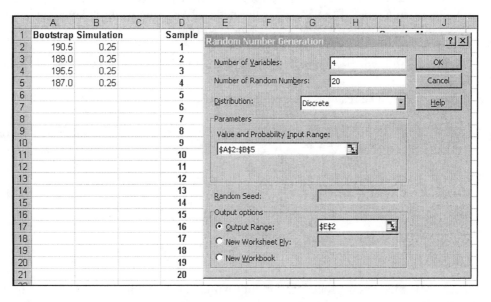

3. Calculate the means of each sample and find the standard deviation of the resulting means. Your answer will be different from the answer shown below because each random sample is different.

	A	B	C	D	E	F	G	H	I	J
1	Bootstrap	Simulation		Sample					Sample Mean	
2	190.5	0.25		1	187	189	190.5	190.5	189.25	
3	189.0	0.25		2	190.5	190.5	187	195.5	190.875	
4	195.5	0.25		3	189	187	189	187	188	
5	187.0	0.25		4	187	195.5	190.5	189	190.5	
6				5	189	195.5	189	189	190.625	
7				6	187	189	195.5	189	190.125	
8				7	190.5	189	190.5	187	189.25	
9				8	189	187	187	195.5	189.625	
10				9	195.5	187	189	190.5	190.5	
11				10	187	189	195.5	190.5	190.5	
12				11	189	190.5	190.5	195.5	191.375	
13				12	187	187	189	190.5	188.375	
14				13	189	190.5	190.5	195.5	191.375	
15				14	189	189	190.5	187	188.875	
16				15	195.5	195.5	190.5	190.5	193	
17				16	195.5	190.5	195.5	190.5	193	
18				17	189	189	187	189	188.5	
19				18	195.5	190.5	190.5	195.5	193	
20				19	189	190.5	190.5	189	189.75	
21				20	187	190.5	190.5	190.5	189.625	
22										
23								Std Dev	1.494769	

4. Construct a histogram of the resulting means using a bin size of 1.

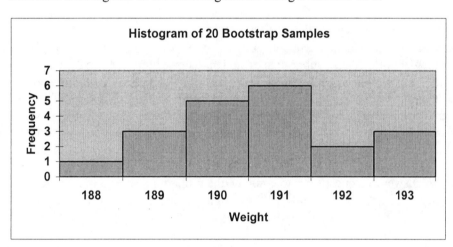

5. If we want a 95% confidence interval, we use the middle 95% of this distribution.

Remarks

The bootstrap method is practical only when you can use a computer to take one thousand or more samples quickly. If this process was repeated for a thousand bootstrap samples, the resulting mean should be a fairly accurate representation of the population mean. More details about the bootstrap method can be found in Chapter 14.

6.2   Tests of Significance

Confidence intervals are appropriate when the goal is to estimate a population parameter such as the mean μ. The second type of inference assesses the evidence provided by the data in favor of some claim about the population.

A significance test compares observed data with a hypothesis that we want to assess. The hypothesis (null) is a statement about the parameters in a population. The results of a test are expressed in terms of a probability that measures how well the data and the hypothesis agree. If the results do not agree, evidence is provided against the null hypothesis.

The z test statistic is calculated to give a standardized value of x: $z = \dfrac{x - \mu}{\sigma}$

The value of the z test statistic is an indication of how far the sample mean is from the null hypothesis. The resulting P-value is calculated as the appropriate proportion greater than or less than the calculated value.

IPS Example 6.7 National Student Loan Survey

One purpose of the National Student Loan Survey described in Example 6.2 is to compare the debt of different subgroups of students. For example, the 525 borrowers who last attended a private four-year college had mean debt of \$21,200, while those who last attended a public four-year college had debt of \$17,100. The difference of \$4,100 is fairly large, but we know that these numbers are estimates of the true means. If we took a different sample, we would get different estimates. Can we conclude that the private four-year students have greater debt than public four-year borrowers from these data?

One way to answer this question is to compute the probability of obtaining a difference as large or larger than the observed \$4,100, assuming that there is no difference in the true means.

The answer is a probability of 0.17. Because this probability is not particularly small, we conclude that observing a difference of \$4,100 is not unusual when the true means are assumed equal. We further conclude that the mean debts for private four-year borrowers and public four-year borrowers are not different.

The statement being tested in a test of significance is called the null hypothesis. The null hypothesis is a statement about a population, expressed in terms of a population parameter, in this case the mean μ. The null hypothesis for this example is stated as:

$$H_0: \mu = 0$$

The statement that we want to test is called the alternative hypothesis. In this example, the alternative hypothesis states that the change in debt is not equal to zero.

The alternative hypothesis is stated as:

$$H_a: \mu \neq 0$$

This is a two-sided problem because it is not obvious whether the mean should be greater than or less than zero. Another possible alternative hypothesis would be a one-sided problem, that is, $\mu > 0$ or $\mu < 0$ if the researcher suspects that the population mean is in only one direction and not in the other.

Calculating Probability for a Particular *x*-Value

1. The standard deviation is determined to be \$3,000, using methods discussed later in this chapter. Calculate the z test statistic using the data given in Example 6.7 by inputting the following formula in an empty cell. The first argument in the function is the x value, the second argument is the population mean, and the third argument is the standard deviation.

   =STANDARDIZE(4100,0,3000)

2. Press **Enter.** The result is the z test statistic.   1.366667

3. The difference of \$4,100 is about one and a third away from the hypothesized value of the population mean.

P-values

A test of significance finds the probability of getting an outcome as extreme or more extreme than the actually observed outcome. In our previous example, we want to calculate the probability of observing a z-value as extreme or more extreme than 1.37. Because this problem was two-sided, the probability needs to be doubled to represent $P(z \leq -1.37 \text{ or } z \geq 1.37)$ where z has a standard normal distribution $N(0,1)$.

4. Calculate the probability of a given z test statistic by inputting the following function in an empty cell.

   =NORMSDIST(1.37)

5. Press **Enter.** The result is the one-sided probability.   0.914656

6. The given probability represents the area to the left of a given z-value. For this example, we want the probability greater than the given value of 1.37. Input the following formula into an empty cell to calculate the area to the right of 1.37:

   =1-0.914656

7. Press **Enter.** The result is the one-sided probability. $\boxed{0.085344}$

8. Input the following formula into an empty cell to double the one-sided probability:

$$\boxed{\text{=2*0.085344}}$$

9. Press **Enter.** The result is the two-sided probability. $\boxed{0.170688}$

Remarks—Interpreting Results for Tests of Significance

We assumed this problem to be a two-sided problem, indicating an uncertainty about whether the result is greater than or less than the accepted value; therefore, the error is doubled. Another way of looking at the two-sided problem is that we add the probability of getting a result as larger or larger than the given value (in this case $\geq \$4{,}100$) to the probability of getting a result as small or smaller than the negative value (in this case, $\leq -\$4{,}100$). Our resulting probability is multiplied by 2, yielding a two-sided probability, or *P*-value, of approximately 0.17, or 17%

Statistical Significance

"Significant" is a statistical term that evaluates the probability of a sample mean result tested against an accepted population mean. Comparison values are expressed by α. Levels of significance are 0.01 or 1%, 0.05 or 5%, and 0.10 or 10%. If a probability or *P*-value is less than the cutoff value for a particular problem (0.05 for 5%), the result is statistically significant at that value. A significant result provides evidence against the accepted value in the null hypothesis and supports the alternative hypothesis.

IPS Example 6.13 Student Debt

The average debt held by a student has risen by $7,500 from 1997 to 2002. Since we would have a prior expectation that the debt would increase over this period because of rising costs of a college education, it is appropriate to use a one-sided alternative in this situation. The hypotheses are

$$H_0: \mu = 0$$

$$H_a: \mu > 0$$

1. Using a standard deviation of $1,900, we calculate the *z* test statistic by inputting the following function into an empty cell. The first argument in the function is the *x*-value, the second argument is the population mean, and the third argument is the standard deviation. $\boxed{\text{=STANDARDIZE(7500,0,1900)}}$

2. Press **Enter.** The result is the *z* test statistic. $\boxed{3.947368}$

3. Calculate the probability of a given *z* test statistic by inputting the following function in an empty cell:

> =NORMSDIST(3.95)

4. Press **Enter.** The result is the one-sided probability to the left of a *z*-value of 3.95.

> 0.999961

5. For this example, we want the probability greater than the given value of 3.95. Input the following formula into an empty cell to calculate the area to the right of 3.95:

> =1-0.999961

6. Press **Enter.** The result is the one-sided probability to the right of a *z*-value of 3.95. If your result is in scientific format (for example, E–05), then click on that cell and select **Format** ⇨ **Cells.** Select the **Number Tab** and select **Number** and five decimal places.

> 0.00004

Remarks

The resulting probability represents about a 4 in 100,000 chance of observing a difference as large or larger than the $7,500 in our sample if the true population difference is zero. This *P*-value tells us that the outcome is extremely rare. This result supports the alternative hypothesis and rejects the null hypothesis.

Tests for a Population Mean

The population mean can be estimated by a sample mean $\bar{x}$. For a particular data set taken from a population that follows a normal distribution, then $\bar{x}$ will be normal and the standard deviation of $\bar{x}$ is represented by $\sigma/\sqrt{n}$. The resulting *z* test statistic is:

$$z = \frac{\bar{x} - \mu_0}{\sigma/\sqrt{n}}$$

Where μ_0 is the hypothesized value of the population mean.

IPS Example 6.14 Blood Pressure for Male Executives

Do middle-aged male executives have different average blood pressure than the general population? The National Center for Health Statistics reports that the mean systolic blood pressure for males 35 to 44 years of age is 128 and the standard deviation in this population is 15. The medical director of a company looks at the medical records of 72 company executives in this age group and find that the mean systolic blood pressure in this sample is $\bar{x} = 126.07$. Is this evidence that executive blood pressures differ from the national average?

1. Using a standard deviation of $15, we calculate the z test statistic by inputting the following function into an empty cell. The first argument in the function is the x-value, the second argument is the population mean, and the third argument is the standard deviation for the sample.

   =STANDARDIZE(126.07,128,15/SQRT(72))

2. Press **Enter.** The result is the z test statistic.   -1.09177

3. Calculate the probability of a given z test statistic by inputting the following function in an empty cell: 0.137467

4. This example is a two-sided problem. Therefore, the resulting probability is doubled.

   =2*0.137467

5. The resulting two-sided probability is   .74934

Remarks

More than 27% of the time, an SRS of size 72 from the male population of interest would have a mean blood pressure at least as far from 128 as that of the executive sample. The observed $\bar{x} = 126.07$ is therefore not good evidence that executives differ from other men.

IPS Example 6.16 Concentrations of Pharmaceutical Samples

The Deely Laboratory analyzes specimens of a pharmaceutical product to determine the concentration of the active ingredient. Such chemical analyses are not perfectly precise. Repeated measurement on the same specimen will give slightly different results. The results of repeated measurements follow a normal distribution quite closely. The analysis procedure has no bias, so that the mean μ of the population of all measurements is the true concentration in the specimen. The standard deviation of this distribution is a property of the analytical procedure and is known to be $\sigma = 0.0068$ grams per liter. The laboratory analyzes each specimen three times and reports the mean result.

The Deely Laboratory has been asked to evaluate the claim that the concentration of the active ingredient in a specimen is 0.86%. The true concentration is the mean μ of the population of repeated analyses.

The hypotheses are: H_0: $\mu = 0.86$

H_a: $\mu \neq 0.86$

Three analyses of one specimen give the following concentrations:

$$0.8403 \qquad 0.8363 \qquad 0.8447$$

1. Find $\bar{x}$ of the sample by using the **AVERAGE** function.

0.8403	
0.8363	
0.8447	
=AVERAGE(C13:C15)	

0.840433

2. The 99% confidence interval for μ is determined by using previous methods.

=CONFIDENCE(.01,0.0068,3)

5. Press **Enter.** The result is the margin of error: $\boxed{0.010113}$

6. The confidence interval can be calculated manually as: 0.8404 +/– 0.0101. The resulting interval is (0.8303, 0.8505).

Remarks

The hypothesized value of $\mu_0 = 0.86$ falls outside of the confidence interval. Therefore, the result is statistically significant. We are 99% confident that μ is not equal to 0.86.

Chapter 7

Inference for Distributions

A normal distribution is described by its center, the mean μ, and its spread, the standard deviation σ. In the previous chapter, methods to estimate and test against the population mean were introduced, assuming that the population standard deviation was known. This chapter continues the topic of inference using one of the most common statistical methods, the t procedures, for the inference about a mean. These procedures no longer assume that the population standard deviation σ is known.

7.1 Inference for the Mean of a Population

The sample mean $\bar{x}$ is used to estimate the unknown population mean μ of a normal distribution in a confidence interval. The sampling distribution depends on the population standard deviation σ and, therefore, does not pose a problem when σ is known. When σ is unknown, we must estimate this parameter, even though we are primarily interested in μ. The sample standard deviation s is used to estimate σ.

The t Distributions

If a simple random sample (SRS) of size n is taken from a population with a normal distribution, with a mean μ and standard deviation σ, the sample mean will have a normal distribution with mean μ and standard deviation $\sigma / \sqrt{n}$. If σ is not known, the spread of the sampling distribution can be estimated by $s / \sqrt{n}$, using the sample standard deviation s. This is known as the **standard error** of the sample mean.

$$SE_{\bar{x}} = \frac{s}{\sqrt{n}}$$

Previously, the z procedures were used when the population standard deviation σ was assumed known. The t procedures are now used when the standard error is substituted for the standard deviation.

For an SRS of size n drawn from a $N(\mu, \sigma)$ population, the one-sample t statistic is defined as

$$t = \frac{\bar{x} - \mu}{s / \sqrt{n}}$$

The t statistic follows the t distribution, which does not have a normal distribution. In fact, there is a different t distribution for each sample size. A particular t distribution is specified by the degrees of freedom, defined as the sample size $n - 1$. The t distribution is symmetric about zero and bell-shaped; however, the spread is greater than the normal distribution due to the variability introduced by using s instead of σ. As the degrees of freedom increase, the t distribution approaches the normal distribution more closely (as s approaches σ, the sample size increases).

The One-Sample t Confidence Interval

The confidence interval for the population mean μ is defined by an SRS of size n drawn from a population with unknown μ and σ:

$$\bar{x} + / - t * \frac{s}{\sqrt{n}}$$

where $t*$ is the critical t statistic for a particular confidence level C (the area between $-t*$ and $t*$ under the curve of a t density with $n - 1$ degrees of freedom equals C).

The margin of error is defined as $t * \dfrac{s}{\sqrt{n}}$.

IPS Example 7.1 Vitamin C in Corn Soy Blend

In fiscal year 1996, the U.S. Agency for International Development purchased 238,300 metric tons of corn soy blend (CSB) for development programs and emergency relief in countries throughout the world. CSB is a highly nutritious, low-cost fortified food that is partially precooked and can be incorporated into different food preparations by the recipients. As part of a study to evaluate appropriate vitamin C levels in this commodity, measurements were taken on samples of CSB produced in a factory.

The following data are the amounts of vitamin C, measured in milligrams per 100 grams (mg/100g) of blend (dry basis), for a random sample of size 8 from a production run:

26	31	23	22	11	22	14	31

We want to find a 95% confidence interval for μ, the mean vitamin C content of the CSB produced during this run.

1. Input the data set into the first column of an empty worksheet or open file **eg07_001** from the **IPS CD-ROM**.

2. The normal critical value t^* is calculated by inputting the following function into a cell to the right of the data set:

    =TINV(0.05,7)

    The probability of 0.0.05 is used for a 95% confidence interval.

3. Press **Enter.** The result is the critical t^* value.   2.364623

4. Calculate $\bar{x}$ by inputting the **AVERAGE** function and press **Enter.**

	26
	31
	23
	22
	11
	22
	14
	31
Average	=AVERAGE(B1:B8)

5. Calculate the sample standard deviation by inputting the **STDEV** function and press **Enter.** The resulting value is 7.19.

	26
	31
	23
	22
	11
	22
	14
	31
Average	22.5
Std Dev	=STDEV(B1:B8)

6. Calculate the standard error by inputting the equation to divide the sample standard deviation by the square root of n ($\frac{s}{\sqrt{n}}$ or $\frac{7.19}{\sqrt{8}}$) and press **Enter.** The resulting value is 2.54.

	26
	31
	23
	22
	11
	22
	14
	31
Average	22.5
Std Dev	7.19
Std Error	=B10/SQRT(8)

7. Calculate the margin of error by inputting the equation to multiply the critical t^* value by the sample standard deviation (2.365*2.54) and press **Enter.** The resulting value is 6.01, rounded to 6.0.

	26	
	31	
	23	
	22	
.	11	
	22	
	14	
	31	
Average	22.5	Crit t*
Std Dev	7.19	2.365
Std Error	2.54	
Margin of Error	=2.365*B11	

8. The results show that the **Confidence Interval** is 22.5 +/– 6.0 or 16.5 to 28.5. We are 95% confident that the mean vitamin C content of the CSB for this run is between 16.5 and 28.5 mg/100g.

9. The condition that the population distribution is normal cannot be checked, however; the sample can be checked for symmetry and outliers by creating a stemplot manually. There are no outliers and the interval is considered valid.

3	1 1
2	2 2 3 6
1	1 4

The One-Sample *t* Test

The *t* test is similar to the *z* test in the previous chapter. However, this time an SRS of size *n* is taken from a population having an unknown mean μ. The null hypothesis is tested: H_0: $\mu = \mu_0$. The standardized *t* statistic is calculated based on the *t* procedures:

$$t = \frac{\bar{x} - \mu}{s/\sqrt{n}}$$

The value of the *t* test statistic is an indication of how far the sample mean is from the null hypothesis. The resulting *P*-value is calculated as the appropriate proportion greater than or less than the calculated value.

IPS Example 7.2 Is the Vitamin C Level Correct?

The specifications for the CSB described in the textbook in Example 7.1 state that the mixture should contain two pounds of vitamin premix for every 2000 pounds of product. These specifications are designed to produce a mean (μ) vitamin C content in the final product of 40 mg/100 g. We can test a null hypothesis that the mean vitamin C content of the production run in the previous example conforms to these specifications.

Specifically, we test

$$H_0: \mu = 40$$

$$H_a: \mu \neq 40$$

1. Using the Example 7.1 data set, $n = 8$, $\bar{x} = 22.5$, and $s = 7.19$. The t test statistic is

$$t = \frac{22.5 - 40}{7.19 / \sqrt{8}}$$

2. Input the following equation into a cell below Example 7.1 data calculations:

	26	
	31	
	23	
	22	
	11	
	22	
	14	
	31	
Average	22.5	Crit t*
Std Dev	7.19	2.365
Std Error	2.54	
Margin of Error	6.01	
t test statistic	=(B9-40)/B11	

3. Press **Enter**. The result is the critical t statistic value.   -6.88

4. If the result is displayed inside parentheses, as (6.88), the cell format can be changed to the format shown above by the following method: Select the cell, select **Format** ⇨ **Cells** from the menu and click on the **Number** tab. Select **Number** for **Category** and click **OK**.

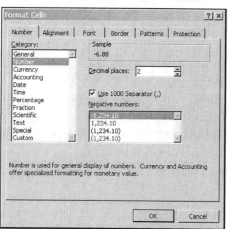

5. The *P*-value associated with the *t* statistic can be calculated using the TDIST function. This function has three arguments. The first is the *t* statistic, input as a positive value of 6.88. Negative values do not work in this function. The second argument is the degrees of freedom, or $n - 1$ or 7. The third argument is the number of tails, which are 2 in this case.

> =TDIST(6.88,7,2)

6. Press **Enter.** The result is the *P*-value for the two-sided *t* test.   0.000236

Remarks

The results indicate that the *P*-value is approximately 0.0002 for the two-sided test. This is evidence against the null hypotheses, leading to the conclusion that the vitamin C content for this run is below the specifications. This also indicates that changes in quality control are necessary to improve the process.

IPS Example 7.3 Has Vitamin C Been Lost in Production?

To test whether the vitamin C content is low, perhaps because vitamin C is lost or destroyed in production, our hypotheses are

$$H_0:\ \mu = 40$$

$$H_a:\ \mu < 40$$

The *t* test statistic has already been calculated in the previous example. The *P*-value for the one-sided test is 0.0001, indicating that the production process has lost some of the vitamin C.

IPS Example 7.4 Diversify or Be Sued

An investor with a stock portfolio worth several hundred thousand dollars sued his broker and brokerage firm because lack of diversification in his portfolio led to poor performance. Table 7.1 in the textbook gives the rates of return for the 39 months that the account was managed by the broker. There are no outliers and we are reasonably confident that the $\bar{x}$ distribution shows no strong skewness. We proceed with our inference based on normal theory.

The arbitration panel compared these returns with the average of the S&P 500 for the same period. Consider the 39 monthly returns as a random sample from the population of monthly returns the brokerage would generate if it managed the account forever. Are these returns compatible with a population mean of $\mu = 0.95\%$, the S&P average?

The testing hypotheses are:

$$H_0:\ \mu = 0.95$$

$$H_a:\ \mu \neq 0.95$$

One-Sample *t* Test Calculation

1. Open the file **ta07_001** from the **IPS CD-ROM.**

2. Calculate $\bar{x}$ by using the AVERAGE function. The resulting value is −1.0985 or −1.1%.

3. Calculate the sample standard deviation and the standard error by inputting the appropriate equations and functions as defined in the previous example.

-1.09846	Mean
5.987785	Std Dev
=A43/SQRT(39)	

4. Calculate the *t* test statistic by inputting the following equation using the previously calculated values:

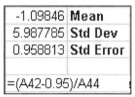

-1.09846	Mean
5.987785	Std Dev
0.958813	Std Error
=(A42-0.95)/A44	

5. Press **Enter.** The result is the *t* test statistic. $\boxed{-2.13646}$

7. The *P*-value associated with the *t* statistic is calculated using the **TDIST** function with the positive equivalent of the *t* statistic, the degrees of freedom, and the number of tails. $\boxed{\text{=TDIST(2.14,38,2)}}$

8. Press **Enter.** The result is the *P*-value for the two-sided *t* test. $\boxed{0.038836}$

Remarks

The resulting two-sided *P*-value of 0.039 indicates that the sample mean of −1.1% is significant at the 5% level and provides evidence against the null hypothesis. That is, the mean return on the client's account differs from that of the stock index.

Matched Pairs *t* Procedures

In a matched pairs study, subjects are matched in pairs because of some common element. The outcomes are compared within each matched pair. Matched pairs can be used for before/after scenarios, to assign two separate treatments randomly to the two subjects in the pair, or when randomization is not possible. The resulting data sets can be analyzed by looking at the difference between the values. The "before" measurements should be subtracted from the "after" measurements. The resulting difference can then be analyzed by using the one-sample t procedures.

IPS Example 7.7 Moon and the Behavior of Some Individuals

Many people believe that the moon influences the behaviors of some individuals. One study of dementia patients in nursing homes recorded various types of disruptive behaviors every day for 12 weeks. Days were classified as moon days if they were in a three-day period centered at the day of the full moon. For each patient, the average number of disruptive behaviors was computed for moon days and for all other days. The data for the 15 subjects in this part of the study for behaviors classified as aggressive are in Table 7.2 in the textbook. The patients in this study are not a random sample of dementia patients. However, we examine their data in the hope that what we find is not unique to this particular group of individuals and applies to other patients who have similar characteristics.

To analyze these data, we first subtract the disruptive behaviors for moon days from the disruptive behaviors for other days. These 15 differences form a single sample. They appear in the "Difference" columns in Table 7.2 in the textbook.

$$H_0: \mu = 0$$

$$H_a: \mu \neq 0$$

where μ is the mean difference in aggressive behaviors on moon days versus other days. The null hypothesis says that aggressive behaviors occur at the same frequency for both types of days, and the alternative hypothesis says that the behaviors on moon days are not the same as on other days.

Matched Pairs—Using a One-Sample t Test

1. Open the file **ta07_002** from the **IPS CD-ROM.**

2. Calculate the sample mean $\bar{x}$ of the difference column, labeled "aggdiff".

3. Calculate the sample standard deviation.

Mean	2.433
Std Dev	1.460

4. The resulting one-sample t statistic is can be input into an empty cell as shown below. Parentheses are necessary in both the numerator and the denominator.

Mean	2.433	
Std Dev	1.460	
t statistic	=(D17-0)/(D18/SQRT(15))	

$$t = \frac{\bar{x} - \mu}{s/\sqrt{n}} = \frac{2.433 - 0}{1.460/\sqrt{15}}$$

5. Press **Enter.** The result is the *t* statistic. | 6.452 |

6. The *P*-value associated with the *t* statistic is calculated using the **TDIST** function using the *t* statistic for the first argument, the degrees of freedom (14) for the second argument, and the number of tails (2) for the third argument.

Mean	2.433
Std Dev	1.460
t statistic	6.452
P-value	=TDIST(D19,14,2)

7. Press **Enter.** The result is the *P*-value for the two-sided *t* test. | 0.000015 |

If the answer is in scientific notation (E–05), change the format of the cell.

Remarks

The resulting *P*-value is very unlikely to occur by chance, therefore there was more aggressive behavior observed on full moon days than on other days.

IPS Example 7.8 Confidence Interval for the Mean Aggressive Behavior

Calculate a 95% confidence interval for the mean difference in aggressive behaviors per day.

Matched Pairs—*t* Confidence Interval

1. Using the difference data set from Example 7.7, calculate the normal critical *t** value by inputting the following function into an empty cell below the data set:

| =TINV(0.05,14) |

The probability of 0.0.05 is used for a 95% confidence interval and 14 for the degrees of freedom.

2. Press **Enter.** The result is the critical *t** value. | 2.145 |

3. Calculate the margin of error as $t * \frac{s}{\sqrt{n}} = 2.145 \frac{1.460}{\sqrt{15}}$ by inputting the following formula into an empty cell below the previous calculation:

Mean	2.433
Std Dev	1.460
t statistic	6.452
P-value	0.000015
t* value	2.145
t statistic	=D21*D18/SQRT(15)

4. Press **Enter.** The result is the margin of error for the difference.   0.81

5. The resulting 95% confidence interval is $\bar{x} +/- t * \dfrac{s}{\sqrt{n}} = 2.43 \pm 0.81 = (1.62, 3.24)$.

Remarks

The estimated average difference is 2.43 aggressive behaviors per day, with a margin of error of 0.81 for 95% confidence.

7.2   Comparing Two Means

Two-sample problems are common in statistical analysis. Two-sample problems are first distinguished from the matched pairs design from the previous chapter. There is no matching of units in the two samples and the two samples can be of different sizes. Each group in the two-sample problem is considered to be from a distinct population or randomly selected from a population and given distinct treatments. The responses in each group are independent of those in the other group. Inference procedures for two-sample data differ from the matched pairs.

Normality of each group can be examined by simple stemplots, histograms, or side-by-side boxplots.

The Two-Sample z Statistic

We are interested in comparing the population means μ_1 and μ_2. If the data are collected from normal populations with known standard deviations σ_1 and σ_2, a variation of the z procedures previously used can be used for inference here. Previously, we compared the sample mean $\bar{x}$ to the population mean μ, dividing by the standard error (SE) σ / n. We are now interested in the differences between two samples and two populations. The resulting two-sample z statistic with a standard normal $N(0, 1)$ sampling distribution is:

$$z = \frac{(\bar{x}_1 - \bar{x}_2) - (\mu_1 - \mu_2)}{\sqrt{\dfrac{\sigma_1^2}{n_1} + \dfrac{\sigma_2^2}{n_2}}}$$

where the standard error (SE) is: $\sqrt{\dfrac{\sigma_1^2}{n_1}+\dfrac{\sigma_2^2}{n_2}}$

The Two-Sample *t* Procedures

The two population standard deviations σ_1 and σ_2 are not generally known. As described in the previous section on *t* procedures, the standard deviation (σ) is replaced by an estimate, the sample standard deviation *s*. The result is the two-sample *t* statistic:

$$t = \frac{(\bar{x}_1 - \bar{x}_2) - (\mu_1 - \mu_2)}{\sqrt{\dfrac{s_1^2}{n_1} + \dfrac{s_2^2}{n_2}}}$$

The statistic does not have a *t* distribution. A *t* distribution replaces a normal distribution only when a single standard deviation σ is replaced by the estimate *s*. We can approximate this distribution by using an approximation for the degrees of freedom that is quite accurate when both sample sizes n_1 and n_2 are five or larger. It is called the Satterthwaite approximation for the degrees of freedom:

$$df = \frac{\left(\dfrac{s_1^2}{n_1} + \dfrac{s_2^2}{n_2}\right)^2}{\dfrac{1}{n_1-1}\left(\dfrac{s_1^2}{n_1}\right)^2 + \dfrac{1}{n_2-1}\left(\dfrac{s_2^2}{n_2}\right)^2}$$

The Two-Sample *t* Significance Test

If an SRS of size n_1 is drawn from a normal population with unknown mean μ_1 and another independent SRS of size n_2 is drawn from another normal population with unknown mean μ_2, the two-sample *t* statistic to test the null hypothesis $H_0: \mu_1 = \mu_2$ is:

$$t = \frac{\bar{x}_1 - \bar{x}_2}{\sqrt{\dfrac{s_1^2}{n_1} + \dfrac{s_2^2}{n_2}}}$$

where the $t(k)$ distribution is used for *P*-values or critical values and the degrees of freedom *k* are the smaller of $n_1 - 1$ and $n_2 - 1$.

IPS Example 7.14 Improving Reading in the Classroom

An educator believes that new directed reading activities in the classroom will help elementary school pupils improve some aspects of their reading ability. She arranges for a third-grade class of 21 students to take part in these activities for an eight-week period. A control classroom of 23 third graders follows the same curriculum without the activities. At the end of the eight weeks, all students are given a Degree of Reading Power (DRP) test, which measures the aspects of reading ability that the treatment is designed to improve. To assess whether the materials improve learning, we test:

$$H_0:\ \mu_1 = \mu_2$$

$$H_a:\ \mu_1 > \mu_2$$

where the treatment group is the first group and the control group is the second group.

A back-to-back stemplot shows a possible outlier in the control group but no deviation form normality that prevents use of t procedures.

1. Open the file **ta07_004** from the **IPS CD-ROM**.

2. Select **Tools** ⇨ **Data Analysis** ⇨ **t-Test: Two-Sample Assuming Unequal Variances** from the menu.

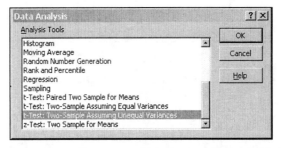

3. Select the **Variable 1 Range** (Treatment Group scores) and the **Variable 2 Range** (Control Group scores). Keep the **Hypothesized Mean Difference** as "0". Select the **Labels** box if you selected labels and keep the **Alpha** value at 0.05. Select an appropriate **Output Range** (upper left of a blank area to the right of the data set) and click **OK**.

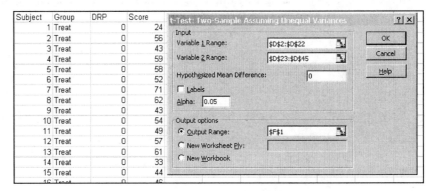

4. Widen the label column of the output. The resulting *P*-value of 1.3% is significant at the 5% level. This method does not calculate the confidence interval.

t-Test: Two-Sample Assuming Unequal Variances		
	Variable 1	Variable 2
Mean	51.47619	41.52174
Variance	121.1619	294.0791
Observations	21	23
Hypothesized Mean Difference	0	
df	38	
t Stat	2.310889	
P(T<=t) one-tail	0.01318	
t Critical one-tail	1.685953	
P(T<=t) two-tail	0.026361	
t Critical two-tail	2.024394	

Remarks

The resulting significant *P*-value of 0.013 uses the Satterthwaite approximation for degrees of freedom. The result is significant at the 5% level but not at the 1% level.

The Two-Sample *t* Confidence Interval

The same concepts used for two-sample *t* significance tests apply in calculating the two-sample *t* confidence intervals. Both samples are SRS from normal populations with unknown means. The confidence interval for $\mu_1 - \mu_2$ is given by:

$$(\bar{x}_1 - \bar{x}_2) \pm t^* \sqrt{\frac{s_1^2}{n_1} + \frac{s_2^2}{n_2}} \quad \text{where the margin or error is} \quad t^* \sqrt{\frac{s_1^2}{n_1} + \frac{s_2^2}{n_2}}$$

The critical *t** value is dependent on the calculation for the degrees of freedom. This method uses the Satterthwaite approximation discussed previously.

IPS Example 7.15 How Much Improvement?

The 95% confidence interval for the mean improvement in the entire population of third graders can be calculated using the data results from the previous example, shown at the top of this page.

1. Input the following formulas into empty cells below the summary data from the previous exercise:

	E	F	G	H	I	J	K
1		Treat	Control		t-Test: Two-Sample Assuming		
2		24	42				
3		56	46			Treat	Control
4		43	43		Mean	51.4761904761905	41.5217391304348
5		59	10		Variance	121.161904761905	294.079051383399
6		58	55		Observations	21	23
7		52	17		Hypothesized Mean Difference	0	
8		71	26		df	38	
9		62	60		t Stat	2.31088919785423	
10		43	62		P(T<=t) one-tail	0.0131803094092657	
11		54	53		t Critical one-tail	1.68595306604402	
12		49	37		P(T<=t) two-tail	0.0263606188185313	
13		57	42		t Critical two-tail	2.02439423446776	
14		61	33				
15		33	37		Diff of Means	=J4-K4	
16		44	41		Margin of Error	=J13*(SQRT((F28^2/J|	
17		46	42		Confidence Interval	=J15-J16	=J15+J16
18		67	19				
19		43	55				
20		49	54				
21		57	28				
22		53	20				
23			48				
24			85				
25							
26		Treat	Control				
27	Mean	=AVERAGE(F2:F22)	=AVERAGE(G2:G24)				
28	Std Dev	=STDEV(F2:F22)	=STDEV(G2:G24)				
29	n	21	23				
30	t statistic	=TTEST(F2:F22,G2:G24,1,3)					

2. The resulting values are shown below and with appropriate significant figures can be stated as $9.96 \pm 8.72 = (1.2, 18.7)$

Diff of Means	9.954451	
Margin of Error	8.720338	
Confidence Interval	1.234114	18.67479

Results

A conservative approach uses 20 for the degrees of freedom, yielding a 95% confidence interval of 9.95 +/– 8.99 or (0.97, 18.94). This approach gives a larger interval than the Satterthwaite approximation of 38 degrees of freedom.

The Pooled Two-Sample *t* Procedures

The one case in which the calculated *t* statistic, used in comparing two means, has exactly a *t* distribution occurs when both normal populations have the same standard deviation, although still unknown.

The resulting confidence level for this special case is defined by

$$(\bar{x}_1 - \bar{x}_2) \pm t^* s_p \sqrt{\frac{1}{n_1} + \frac{1}{n_2}}$$

where s_p is defined as the **pooled estimator** of σ and is defined by the square root of

$$s_p^2 = \frac{(n_1 - 1)s_1^2 + (n_2 - 1)s_2^2}{n_1 + n_2 - 2}$$

The corresponding **pooled two-sample *t* statistic** is:

$$t = \frac{\bar{x}_1 - \bar{x}_2}{s_p \sqrt{\dfrac{1}{n_1} + \dfrac{1}{n_2}}}$$

IPS Example 7.19 Does Calcium Reduce Blood Pressure?

Does increasing the amount of calcium in our diet reduce blood pressure? Examination of a large sample of people revealed a relationship between calcium intake and blood pressure, but such observational studies do not establish causation. Animal experiments then showed that calcium supplements do reduce blood pressure in rats, justifying an experiment with human subjects. A randomized comparative experiment gave one group of 10 black men a calcium supplement for 12 weeks. The control group of 11 black men received a placebo that appeared identical. Table 7.5 in the textbook gives the seated systolic (heart contracted) blood pressure for all subjects at the beginning and end of the 12-week period, in millimeters (mm) of mercury. Because the researchers were interested in decreasing blood pressure, Table 7.5 also shows the decrease for each subject. An increase appears as a negative entry.

Take Group 1 to be the calcium group and Group 2 to be the placebo group. The evidence that calcium lowers blood pressure more than a placebo is assessed by testing

$$H_0\text{: } \mu_1 = \mu_2$$
$$H_a\text{: } \mu_1 > \mu_2$$

1. Open the file **ta07_005** from the **IPS CD-ROM**.

2. Create summary data Select **Tools** $\Rightarrow$ **Data Analysis** $\Rightarrow$ **t-Test: Two-Sample Assuming Equal Variances** from the menu.

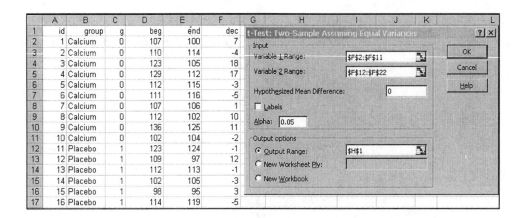

3. Select the **Variable 1 Range** (Calcium Group decreased values) and the **Variable 2 Range** (Control Group decreased values). Keep the **Hypothesized Mean Difference** as "0". Select the **Labels** box if you selected labels and keep the **Alpha** value at 0.05. Select an appropriate **Output Range** (upper left of a blank area to the right of the data set) and click **OK.**

	A	B	C	D	E	F	G	H	I	J	K
1	id	group	g	beg	end	dec		t-Test: Two-Sample Assuming Equal Variances			
2	1	Calcium	0	107	100	7					
3	2	Calcium	0	110	114	-4			Variable 1	Variable 2	
4	3	Calcium	0	123	105	18		Mean	5	-0.27273	
5	4	Calcium	0	129	112	17		Variance	76.44444	34.81818	
6	5	Calcium	0	112	115	-3		Observations	10	11	
7	6	Calcium	0	111	116	-5		Pooled Variance	54.53589		
8	7	Calcium	0	107	106	1		Hypothesized Mean	0		
9	8	Calcium	0	112	102	10		df	19		
10	9	Calcium	0	136	125	11		t Stat	1.634108		
11	10	Calcium	0	102	104	-2		P(T<=t) one-tail	0.059348		
12	11	Placebo	1	123	124	-1		t Critical one-tail	1.729131		
13	12	Placebo	1	109	97	12		P(T<=t) two-tail	0.118697		
14	13	Placebo	1	112	113	-1		t Critical two-tail	2.093025		
15	14	Placebo	1	102	105	-3					
16	15	Placebo	1	98	95	3					
17	16	Placebo	1	114	119	-5					
18	17	Placebo	1	119	114	5					
19	18	Placebo	1	112	114	2					
20	19	Placebo	1	110	121	-11					
21	20	Placebo	1	117	118	-1					
22	21	Placebo	1	130	133	-3					
23											
24					St Dev C	8.743251					
25					St Dev P	5.900693					

4. Calculate the sample standard deviations of both groups. These values are close enough to assume equal standard deviations.

Results

The pooled two-sample variance is shown as 54.536 and the pooled two-sample *t* statistic is 1.634. The resulting one-sided probability is 0.059, which is slightly over the limit for significance at the 5% level.

Two-Sample Pooled Confidence Interval

IPS Example 7.21 90% Confidence Interval for Effect of Calcium

We estimate that the effect of calcium supplementation is the difference between the sample means of the calcium and the placebo groups, to be:

$\bar{x}_1 - \bar{x}_2 = (5.00 - (-0.273)) = 5.273mm.$

1. The critical *t*-value for a 90% level for a one-sided test is shown on the summary data in the previous exercise as 1.634.

t-Test: Two-Sample Assuming Equal Variances		
	Variable 1	Variable 2
Mean	5	-0.27273
Variance	76.44444	34.81818
Observations	10	11
Pooled Variance	54.53589	
Hypothesized Mean	0	
df	19	
t Stat	1.634108	
P(T<=t) one-tail	0.059348	
t Critical one-tail	1.729131	
P(T<=t) two-tail	0.118697	
t Critical two-tail	2.093025	

2. The pooled variance is shown as 54.536. Find the square root of that value, using the SQRT function to calculate s_p. The resulting value is 7.385.

3. Solve the confidence interval equation $(\bar{x}_1 - \bar{x}_2) \pm t * s_p \sqrt{\dfrac{1}{n_1} + \dfrac{1}{n_2}}$ by substituting the known and calculated values yielding $5.000 - (-0.2727) \pm 1.729 * 7.385 \sqrt{\dfrac{1}{10} + \dfrac{1}{11}}$ or

    $5.273 \pm 12.7687 \sqrt{\dfrac{1}{10} + \dfrac{1}{11}}$ or $5.273 \pm 12.7687 \sqrt{\dfrac{1}{10} + \dfrac{1}{11}}$ or by inputting the equation for the margin of error.

    =1.729*7.3848*SQRT((1/10)+(1/11))

6. Press **Enter.** The result is the margin of error for the difference.   5.57888

7. The resulting 90% confidence interval is $5.273 \pm 5.5789 = (-0.306, 10.852)$.

Remarks

We are 90% confident that the difference in means is in the interval $(-0.306, 10.852)$.

Chapter 8

Inference for Proportions

Some statistical studies record categorical variables that can be measured as counts or percents. The population parameters of interest are the population proportions in the separate categories. We can analyze a single proportion or compare two proportions. The methods of analysis are similar to methods used for inference about means discussed in Chapter 7. Both methods of inference are based on sampling distributions that are approximately normal.

8.1   Inference for a Single Proportion

The statistic used to estimate the unknown parameter is the sample proportion. An SRS of size n is chosen from a large population that contains population proportion p of "successes." Define $\hat{p}$ as the sample proportion of successes:

$$\hat{p} = \frac{count\ of\ successes\ in\ the\ sample}{n} = \frac{X}{n}$$

where X is a discrete random variable representing the count of "Yes" answers in a sample. X can take values such as 0, 1/100, 2/100…, 99/100, or 1.

For a large n, the sampling distribution of $\hat{p}$ becomes approximately Normal with a mean of p and a standard deviation defined by

$$\sqrt{\frac{p(1-p)}{n}}$$

Modern computer studies have shown that confidence intervals based on this statistic, even for large samples, can be inaccurate. Studies have also shown that the estimate is improved by moving the sample proportion $\hat{p}$ slightly away from zero and one. A simple adjustment can be made by adding four additional observations, two of which are successes and two of which are failures. The new sample size is $n + 4$ and the count of successes is $X + 2$. The new estimator of the population proportion is

$$\tilde{p} = \frac{X+2}{n+4}$$

This is called the **Wilson estimate** of the population proportion. The resulting standard error of $\hat{p}$ is:

$$SE_{\tilde{p}} = \sqrt{\frac{\tilde{p}(1-\tilde{p})}{n+4}}$$

An approximate level C confidence interval for p is:

$$\tilde{p} \pm z^* SE_{\tilde{p}}$$

z^* is defined as the critical z-value defined by the confidence level C (the area between—z^* and z^* using a normal curve equals C). This interval can be used when the sample size is at least $n = 5$ and the confidence level is 90%, 95%, or 99%.

The resulting margin of error is

$$m = z^* SE_{\tilde{p}}$$

Large-Sample Confidence Interval for a Single Proportion

The unknown population proportion p is estimated by the sample proportion $\hat{p} = \dfrac{X}{n}$ where X is the number of successes. If the sample size n is large enough, $\hat{p}$ has an approximately normal distribution, with mean $\mu_{\hat{p}} = p$ and standard deviation $\sigma_{\hat{p}} = \sqrt{p(1-p)/n}$. This means that, approximately 95% of the time, $\hat{p}$ will be within $2\sqrt{p(1-p)/n}$ of the unknown population proportion p. To estimate the standard deviation using the data, we replace p in the formula by the sample proportion $\hat{p}$. We also use the term **standard error** for the standard deviation of a statistic that is estimated from data.

The standard error of $\hat{p}$ is $SE_{\hat{p}} = \sqrt{\dfrac{\hat{p}(1-\hat{p})}{n}}$, and the margin of error for confidence level C is $m = z^* SE_{\hat{p}}$, where z^* is the critical z-value.

IPS Example 8.1 Alcohol Abuse on Campus

Alcohol abuse has been described by college presidents as the number one problem on campus, and it is an important cause of death in young adults. How common is it? A survey of 17,096 students in U.S. four-year colleges collected information on drinking behavior and alcohol-related problems. The researchers defined "frequent binge drinking" as having five or more drinks in a row three or more times in the past two weeks. According to this definition, 3,314 students were classified as frequent binge drinkers.

1. Input the summary data into an empty worksheet.

2. First, calculate the Wilson estimate of the proportion of drinkers by solving the following formula:

$$\hat{p} = \frac{3,314}{17,096} = 0.194$$

	A	B
1	Summary Data	
2		
3	X	3314
4	n	17096
5		
6	p-hat	=B3/B4
7	SE	=SQRT((B6*(1-B6))/B4)
8	z*	=NORMSINV(0.975)
9	Margin of Error	=B8*B7
10	Lower Limit	=B6-B9
11	Upper Limit	=B6+B9

3. Compute the standard error, critical z-value, the margin of error and the resulting lower and upper limits of the confidence interval.

	A	B
1	Summary Data	
2		
3	X	3314
4	n	17096
5		
6	p-hat	0.194
7	SE	0.00302337
8	z*	1.95996108
9	Margin of Error	0.00592568
10	Lower Limit	0.188
11	Upper Limit	0.200

Remarks

Using the **Wilson Estimate**, we are 95% confident that between 18.8% and 20.0% of the college students are frequent binge drinkers.

Plus Four Confidence Interval for a Single Proportion

It has been shown that confidence intervals based on the large-sample approach can be inaccurate. An adjustment is made by adding two successes and two failures yielding a new sample size of $n + 4$ and the count of successes of $X + 2$. The resulting population proportion based on the **plus four** rule is:

$$\tilde{p} = \frac{X + 2}{n + 4}$$

The standard error of $\tilde{p}$ is $SE_{\tilde{p}} = \sqrt{\dfrac{\tilde{p}(1-\tilde{p})}{n+4}}$, and the margin of error for confidence level C is $m = z * SE_{\tilde{p}}$, where $z*$ is the critical z-value.

The confidence interval for p is $\tilde{p} \pm m$ and can be used for samples sizes at least $n = 10$.

IPS Example 8.2 Health Benefits Associated with Soy

Research has shown that there are many health benefits associated with a diet that contains soy foods. Substances in soy called isoflavones are known to be responsible for these benefits. When soy foods are consumed, some subjects produce a chemical called equol and it is thought that production of equol is a factor in the health benefits of a soy diet. Unfortunately, not all people are equol producers. A nutrition researcher planning some bone health experiments would like to include some equol producers and some non producers among her subjects. A preliminary sample of 12 female subjects were measured and it was found that 4 were equol producers. We would like to estimate the proportion of equol producers in the population from which this researcher will draw her subjects.

1. Input the summary data into an empty worksheet.

2. First, calculate the plus four estimate of the proportion of equol producers by solving the following formula: $\hat{p} = \dfrac{4+2}{12+4} = \dfrac{6}{16} = 0.375$

	A	B
1	Summary Data	
2		
3	*X*	4
4	*X* + 2	=+B3+2
5	*n*	=12
6	*n+4*	=B5+4
7		
8	p-hat	=B4/B6
9	SE	=SQRT((B8*(1-B8))/B6)
10	z*	=NORMSINV(0.975)
11	Margin of Error	=B10*B9
12	Lower Limit	=B8-B11
13	Upper Limit	=B8+B11

3. Compute the standard error, critical z-value, the margin of error and the resulting lower and upper limits of the confidence interval.

p-hat	0.375
SE	0.1210307
z*	1.9599611
Margin of Error	0.2372155
Lower Limit	0.138
Upper Limit	0.612

Remarks

Using the **plus four estimate**, we are 95% confident that between 13.8% and 61.2% of women from this population are equol producers.

Large-Sample Significance Test for a Single Proportion

A significance test requires a null hypothesis value to test against, which can be called p_0. When calculating the *P*-value, we test this value as the accepted value. The resulting null hypothesis is

$$H_0: p = p_0$$

The alternative hypothesis can test $p > p_0$, $p < p_0$ as one-sided tests or $p \neq p_0$ as a two-sided test.

If an SRS of size n is chosen from a large population with unknown proportion p of successes, the hypotheses can be tested by computing the following z statistic:

$$z = \frac{\hat{p} - p_0}{\sqrt{\dfrac{p_0(1 - p_0)}{n}}}$$

This test can be used when the expected number of successes np_0 and the expected number of failures $n(1 - p_0)$ are both greater than 10.

IPS Example 8.3 Work Stress

According to the National Institute for Occupational Safety and Health, job stress poses a major threat to the health of workers. A national survey of restaurant employees found that 75% said that work stress had a negative impact on their personal lives. A sample of 100 employees of a restaurant chain finds that 68 answer "Yes" when asked, "Does work stress have a negative impact on your personal life?" Is this good reason to think that the proportion of all employees of this chain who would say "Yes" differs from the national proportion $p_0 = 0.75$?

$$H_o: p = 0.75$$

$$H_a: p \neq 0.75$$

1. Input the summary data into an empty worksheet.

2. Compute the standard error, z-value, and the resulting *P*-value. Multiply the *P*-value by two for a two-sided probability.

	A	B
1	**Summary Data**	
2		
3	X	68
4	*n*	100
5	p_0	0.75
6	p-hat	=B3/B4
7	SE	=SQRT((B6*(1-B6))/B4)
8	z	=(B6-B5)/SQRT((B5*(1-B5))/B4)
9	*P*-value	=NORMSDIST(B8)
10	*P*-value*2	=2*B9

3. The results show two-sided *P*-value of 10.6%, which means that the restaurant chain survey is in line with the national results. The result is not significant.

	A	B
1	**Summary Data**	
2		
3	X	68
4	*n*	100
5	p_0	0.75
6	p-hat	0.680
7	SE	0.04665
8	z	-1.61658
9	*P*-value	0.05298
10	*P*-value*2	0.106

4. A 95% confidence interval can be calculated using methods previously described.

12	z*	=NORMSINV(0.975)
13	margin of error	=B12*B7
14	Lower Limit	=B6-B13
15	Upper Limit	=B6+B13

12	z*	1.95996
13	margin of error	0.09143
14	Lower Limit	0.589
15	Upper Limit	0.771

Remarks

We are 95% confident that between 59% and 77% of the restaurant chain's employees feel that work stress is damaging to their personal lives.

Choosing a Sample Size

The margin of error for the confidence interval for a population proportion is

$$m = z * SE_{\hat{p}} = z * \sqrt{\frac{\hat{p}(1-\hat{p})}{n}}$$

where $z*$ is obtained from the confidence level.

The value of $\hat{p}$ is not known until we gather data, so we will estimate it by a guess value $p*$. The value of $p*$ could be a sample estimate from a similar or previous study or use a value of 0.5. The margin of error will be the largest when $\hat{p} = 0.5$, which will provide a conservative estimate for the resulting sample size. Once $p*$ is chosen and we have decided on the desired margin of error m, the sample size can be calculated as follows:

$$n = \left(\frac{z*}{m}\right)^2 p*(1-p*)$$

If $p* = 0.5$, the sample size required is given by

$$n = \left(\frac{z*}{2m}\right)^2$$

The value of n calculated in this method is not very sensitive to the choice of $p*$ as long as $p*$ is not too far from 0.5. However, if the sample turns out to have a $\hat{p}$ smaller than about 0.3 or larger than about 0.7, the sample size based on $p* = 0.5$ may be much larger than needed.

IPS Example 8.6 Planning a Sample of Customers

Your company has received complaints about its customer support service. You intend to hire a consulting company to carry out a sample survey of customers. Before contacting the consultant, you want some idea of the sample size you will have to pay for. One critical question is the degree of satisfaction with your customer service, measured on a five-point scale. You want to estimate the proportion p of your customers who are satisfied (that is, who choose either "satisfied" or "very satisfied," the two highest levels on the five-point scale).

You want to estimate p with 95% confidence, and a margin of error less than or equal to 3%, or 0.03. For planning purposes, you can use $p* = 0.5$.

1. Input the summary data into an empty worksheet. Compute the critical $z*$-value and the resulting sample size.

	A	B
1	Summary Data	
2		
3	*p*.	0.5
4	*z**	=NORMSINV(0.975)
5		
6	margin of error	0.03
7	Sample Size	=(B4/(2*B6))^2

2. The resulting sample size of 1067.1 rounds up to 1068 to ensure that the margin of error is satisfied.

	A	B
1	Summary Data	
2		
3	*p*.	0.5
4	*z**	1.95996
5		
6	margin of error	0.03
7	Sample Size	1067.069

3. If this method is repeated for a margin of error of 2%, the required sample size more than doubles to almost 2400. The margin of error can be input as either a percent or a decimal value.

	A	B
1	Summary Data	
2		
3	*p*.	0.5
4	*z**	1.9599611
5		
6	margin of error	0.02
7	Sample Size	2400.905

Remarks

As the desired margin of error decreases, the sample size increases. If $\hat{p}$ is difficult to estimate, several values of $\hat{p}$ and their resulting margin of error can be calculated in order to determine the most reasonable choice for the sample size.

IPS Example 8.7 Factors That Influence Sample Size

The Division of Recreational Sports (Rec Sports) at a major university is responsible for offering comprehensive recreational programs, services, and facilities to the students. They are continually examining their programs to determine how well they are meeting the needs of their students. Rec Sports is considering adding some new programs and would like to know how much interest there would be in a new exercise program based on the Pilate method. They will take a survey of undergraduate students. In the past, they sent emails to all undergraduate students, asking them to complete a short survey. The response rate obtained in this way was about 5%. This time, they will send emails to a simple random sample of the students and will follow up with additional emails and eventually a phone call to get a higher response rate. Because of the work involved with the follow-up and limited staff, they would like to use a sample size of about 200. Would the sample size of $n = 200$ be adequate to provide Rec Sports with the information they need?

Calculate the margins of error for 95% confidence intervals for various values of $\hat{p}$.

1. Input the summary data into a worksheet and the formulas to calculate the standard error and the resulting margin of error for several values of $\hat{p}$. The formulas are copied down to use the different values of $\hat{p}$. The cell reference for z^* has to be made an absolute cell reference before it can be successfully copied down.

	A	B	C
1		**Summary Data**	
2			
3	z^*	=NORMSINV(0.975)	
4			
5		p_{est}	m
6		0.05	=$B$3*SQRT((B6*(1-B6))/200)
7		0.1	=$B$3*SQRT((B7*(1-B7))/200)
8		0.15	=$B$3*SQRT((B8*(1-B8))/200)
9		0.2	=$B$3*SQRT((B9*(1-B9))/200)
10		0.25	=$B$3*SQRT((B10*(1-B10))/200)
11		0.3	=$B$3*SQRT((B11*(1-B11))/200)
12		0.35	=$B$3*SQRT((B12*(1-B12))/200)
13		0.4	=$B$3*SQRT((B13*(1-B13))/200)
14		0.45	=$B$3*SQRT((B14*(1-B14))/200)
15		0.5	=$B$3*SQRT((B15*(1-B15))/200)
16		0.55	=$B$3*SQRT((B16*(1-B16))/200)
17		0.6	=$B$3*SQRT((B17*(1-B17))/200)
18		0.65	=$B$3*SQRT((B18*(1-B18))/200)
19		0.7	=$B$3*SQRT((B19*(1-B19))/200)
20		0.75	=$B$3*SQRT((B20*(1-B20))/200)
21		0.8	=$B$3*SQRT((B21*(1-B21))/200)
22		0.85	=$B$3*SQRT((B22*(1-B22))/200)
23		0.9	=$B$3*SQRT((B23*(1-B23))/200)
24		0.95	=$B$3*SQRT((B24*(1-B24))/200)

2. The resulting margins of error were acceptable to Rec Sports. They decided to use a sample size of 500 in the survey.

	A	B	C
1		**Summary Data**	
2			
3	z^*	1.959961	
4			
5		p_{est}	m
6		0.05	0.030
7		0.10	0.042
8		0.15	0.049
9		0.20	0.055
10		0.25	0.060
11		0.30	0.064
12		0.35	0.066
13		0.40	0.068
14		0.45	0.069
15		0.50	0.069
16		0.55	0.069
17		0.60	0.068
18		0.65	0.066
19		0.70	0.064
20		0.75	0.060
21		0.80	0.055
22		0.85	0.049
23		0.90	0.042
24		0.95	0.030

8.2   Comparing Two Proportions

The proportions of two groups can be compared if independent SRSs are chosen from separate populations. To compare the two unknown population proportions, we start by comparing the difference between the two sample proportions of successes. The estimate of the difference in the population proportions is $D = \hat{p}_1 - \hat{p}_2$.

As both sample sizes increase, the sampling distribution of the difference becomes approximately normal. The standard error of D is:

$$SE_D = \sqrt{\frac{\hat{p}_1(1-\hat{p}_1)}{n_1} + \frac{\hat{p}_2(1-\hat{p}_2)}{n_2}}$$

The margin of error for a confidence level C is $m = z^* SE_D$. The resulting confidence interval for $p_1 - p_2$ is $D \pm m$. This method is used when the number of successes and the number of failures in both samples are all at least 10.

IPS Example 8.9 Confidence Interval for Proportions of Binge Drinkers

Let's find a 95% confidence interval for the difference between the proportions of men and of women who are frequent binge drinkers.

1. The data used are in Example 8.1. Input the summary data into a worksheet and the formulas to calculate the standard error and the resulting margin of error for several values of $\hat{p}$.

	A	B	C	D
1	Summary Data			
2				
3	Population	n	X	p-hat
4	1 (men)	7180	1630	=C4/B4
5	2 (women)	9916	1684	=C5/B5
6	Total	=SUM(B4:B5)	=SUM(C4:C5)	=C6/B6

2. The resulting values are shown below.

	A	B	C	D
1	Summary Data			
2				
3	Population	n	X	p-hat
4	1 (men)	7,180	1,630	0.227
5	2 (women)	9,916	1,684	0.170
6	Total	17,096	3,314	0.194

3. Input the equations to calculate the standard error, margin of error, and the resulting lower and upper limits for the 95% confidence interval.

	A	B	C	D
1	**Summary Data**			
2				
3	*Population*	*n*	*X*	*p-hat*
4	1 (men)	7180	1630	=C4/B4
5	2 (women)	9916	1684	=C5/B5
6	Total	=SUM(B4:B5)	=SUM(C4:C5)	=C6/B6
7				
8	z*	=NORMSINV(0.975)		
9	D	=D4-D5		
10	Standard Error	=SQRT((D4*(1-D4))/B4+(D5*(1-D5))/B5)		
11	margin of error	=B8*B10		
12	Lower Limit	=B9-B11		
13	Upper Limit	=B9+B11		

4. The resulting values indicate that with 95% confidence we can say that the difference in the proportions is between 0.045 and 0.069.

	A	B	C	D
1	**Summary Data**			
2				
3	*Population*	*n*	*X*	*p-hat*
4	1 (men)	7,180	1,630	0.227
5	2 (women)	9,916	1,684	0.170
6	Total	17,096	3,314	0.194
7				
8	z*	1.960		
9	D	0.057		
10	Standard Error	0.0062		
11	margin of error	0.0122		
12	Lower Limit	0.045		
13	Upper Limit	0.069		

Plus Four Confidence Interval for a Difference in Proportions

We will use the same Wilson approximation used for one proportion in order to improve the accuracy of confidence intervals. The **plus four** estimate of the difference in the population proportions is $\tilde{D} = \tilde{p}_1 - \tilde{p}_2$, where

$$\tilde{p}_1 = \frac{X_1 + 1}{n_1 + 2} \quad \text{and} \quad \tilde{p}_2 = \frac{X_2 + 1}{n_2 + 2}$$

The standard error of the difference is defined by

$$SE_{\tilde{D}} = \sqrt{\frac{\tilde{p}_1(1 - \tilde{p}_1)}{n_1 + 2} + \frac{\tilde{p}_2(1 - \tilde{p}_2)}{n_2 + 2}}$$

An approximate level C confidence interval for $p_1 - p_2$ is

$$(\tilde{p}_1 - \tilde{p}_2) \pm z^* SE_{\tilde{D}}$$

where z^* is defined as the critical z-value defined by the confidence level C (the area between $-z^*$ and z^* using a normal curve equals C).

The resulting margin of error is

$$m = z^* SE_{\tilde{D}}$$

An approximate level C confidence interval for $p_1 - p_2$ is rephrased as $\tilde{D} \pm m$.

This method can be used when both sample sizes are at least 10 and the confidence level is 90%, 95%, or 99%.

IPS Example 8.10 Confidence Interval for Differences in Gender Proportions

In studies that look for a difference between genders, a major concern is whether or not apparent differences are due to other variables that are associated with gender. Because boys mature more slowly than girls, a study of adolescents that compares boys and girls of the same age may confuse a gender effect with an effect of sexual maturity. The "Tanner score" is a commonly used measure of sexual maturity. Subjects are asked to determine their score by placing a mark next to a rough drawing of an individual at their level of sexual maturity. There are five drawings, so the score is an integer between 1 and 5.

A pilot study included 12 girls and 12 boys from a population that will be used for a large experiment. Four of the boys and three of the girls had Tanner scores of 4 or 5, a fairly high level of sexual maturity. Find a 95% confidence interval for the difference between the proportions of boys and of girls who have high (4 or 5) Tanner scores in this population. The large sample approach is not recommended. However, the sample sizes are both at least 5, so the plus four method is appropriate.

1. Input the summary data into a worksheet and the formulas to calculate the standard error, the resulting margin of error, and the lower and upper limits for the 95% confidence interval.

	A	B	C	D	E	F
1	**Summary Data**					
2						
3	*Population*	*n*	*n + 2*	*X*	*X + 1*	*p-hat*
4	1 (boys)	12	=B4+2	4	=D4+1	=E4/C4
5	2 (girls)	12	=B5+2	3	=D5+1	=E5/C5
6	Total	=SUM(B4:B5)	=SUM(C4:C5)	=SUM(D4:D5)	=SUM(E4:E5)	=E6/C6
7						
8	z*	=NORMSINV(0.975)				
9	D	=F4-F5				
10	Standard Error	=SQRT((F4*(1-F4))/C4+(F5*(1-F5))/C5)				
11	margin of error	=B8*B10				
12	Lower Limit	=B9-B11				
13	Upper Limit	=B9+B11				

2. The resulting values are shown below.

	A	B	C	D	E	F
1	Summary Data					
2						
3	*Population*	*n*	*n + 2*	*X*	*X + 1*	*p*-hat
4	1 (boys)	12	14	4	5	0.357
5	2 (girls)	12	14	3	4	0.286
6	Total	24	28	7	9	0.321
7						
8	z*	1.960				
9	D	0.071				
10	Standard Error	0.1760				
11	margin of error	0.3450				
12	Lower Limit	-0.274				
13	Upper Limit	0.416				

Remarks

With 95% confidence, we can say that the difference in the proportions is between –0.274 and 0.416.

Significance Test for Two Proportions

Assume an SRS of size n_1 is taken from a large population having proportion p_1 of successes and an independent SRS of size n_2 is taken from another population having proportion p_2 of successes. When calculating the *P*-value, we compare p_1 to p_2. The resulting null hypothesis is:

$$H_0: p_1 = p_2$$

The hypotheses are tested by computing the following *z* statistic:

$$z = \frac{\hat{p}_1 - \hat{p}_2}{SE_{Dp}}$$

where the pooled standard error is

$$SE_{Dp} = \sqrt{\hat{p}(1 - \hat{p})\left(\frac{1}{n_1} + \frac{1}{n_2}\right)}$$

The pooled standard error represents the common proportion of successes:

$$\hat{p} = \frac{X_1 + X_2}{n_1 + n_2}$$

IPS Example 8.11 Gender Differences in Frequent Binge Drinkers

Are men and women college students equally likely to be frequent binge drinkers? We examine the survey data in Example 8.8 to answer this question.

1. Input the summary data into a worksheet and the formulas to calculate the standard error and the resulting margin of error for several values of $\hat{p}$.

	A	B	C	D
1	Summary Data			
2				
3	*Population*	*n*	*X*	*p*-hat
4	1 (men)	7180	1630	=C4/B4
5	2 (women)	9916	1684	=C5/B5
6	Total	=SUM(B4:B5)	=SUM(C4:C5)	=C6/B6
7				
8	Standard Error	=SQRT((D6*(1-D6))*((1/B4)+(1/B5)))		
9	z	=(D4-D5)/B8		
10	*P*-value	=NORMSDIST(-B9)		
11	*P*-value * 2	=2*B10		

2. The *P*-value was calculated using a negative *z*-value to get the tail probability. Results indicate that the difference in the proportions between men and women binge drinkers is statistically significant at the 1% level.

	A	B	C	D
1	Summary Data			
2				
3	*Population*	*n*	*X*	*p*-hat
4	1 (men)	7,180	1,630	0.227
5	2 (women)	9,916	1,684	0.170
6	Total	17,096	3,314	0.194
7				
8	Standard Error	0.006126		
9	z	9.3366		
10	*P*-value	0.0000000		
11	*P*-value * 2	0.000000		

Chapter 9

Inference for Two-Way Tables

In Chapter 8, we compared samples by analyzing the number of observations (n), classified as "successes" (X), analyzing inference about proportions for both one and two samples. We now turn to comparing two or more populations to determine whether two categorical variables are independent. We begin with data analysis methods for two-way tables.

9.1   Data Analysis for Two-Way Tables

In Chapter 8, inference for two proportions is examined using summary data in a table of the form below, using Example 8.9 data.

Population	n	X
1 (men)	7,180	1,630
2 (women)	9,916	1,684
Total	17,096	3,314

In this chapter, we will summarize data in a different way by using a two-way table, recording the counts for all possible outcomes. The following is a two-way table identifying the actual counts of women and men who were or were not binge drinkers.

Frequent Binge Drinker	Gender		Total
	Men	Women	
Yes	1,630	1,684	3,314
No	5,550	8,232	13,782
Total	7,180	9,916	17,096

This table format rearranges the information previously gathered and displays it in terms of two categorical variables—gender and whether or not the person was a frequent binge drinker. This is a two-way table because there are two columns (defining Women and Men) and two rows (defining Yes and No as Frequent Binge Drinker). One of these variables could be considered as the explanatory variable (Gender) and the other variable as the response variable (Binge Drinker). Two-way tables were previously discussed in the textbook in Chapter 2. In this chapter, both of the variables are classified as categorical. Our hypotheses will involve whether there exists a relationship between the row variable and the column variable. For example, is there a relationship between someone's Gender and being a Frequent Binge Drinker?

Analysis of two-way tables is made possible by using Excel or statistical software to carry out the calculations. To describe the relationship between categorical variables, we compute and compare percentages. A count in each cell can be viewed as a percent of the total, a percent of a row total (called the joint distribution), or a percent of a column total (called the conditional distribution). It is up to the researcher to decide which percentages are important for each problem. This table also includes the marginal totals and the grand total, which represents the sum of both the rows and the columns.

Joint Distribution

IPS Example 9.3 Frequent Binge Drinking and Gender

The counts in the table shown on the previous page represent the sample data for this study. As stated above, the two categorical variables in this case are whether or not a person is a frequent binge drinker and whether or not that person is a man or a woman. We are interested in whether or not gender has an influence on whether or not someone is a frequent binge drinker. Therefore, gender would be the explanatory variable, and whether or not a person was a frequent binge drinker would be the response variable. Not all two-way tables have an explanatory and response variable. It is possible to simply display data categorized in this fashion without regard to a specific relationship between the variables.

A cell is defined as the intersection of a row and a column. A two-way table with r rows and c columns contains $r \times c$ cells.

1. We are going to calculate the joint distribution of binge drinking and gender. First, input the tables shown below into an empty worksheet.

	A	B	C	D
1		Gender		
2	Frequent Binge Drinker	Men	Women	Total
3	Yes	1,630	1,684	3,314
4	No	5,550	8,232	13,782
5	Total	7,180	9,916	17,096
6				
7	Joint Distribution			
8	Frequent Binge Drinker	Men	Women	
9	Yes			
10	No			

2. Input the formulas shown below into the second table. The formulas divide each count by the total sample size. This is the **joint distribution.** The reference for the total should be absolute (denoted by the $ before both the column letter and row number) if the formula is copied.

	A	B	C	D
1		Gender		
2	Frequent Binge Drinker	Men	Women	Total
3	Yes	1,630	1,684	3,314
4	No	5,550	8,232	13,782
5	Total	7,180	9,916	17,096
6				
7	Joint Distribution			
8	Frequent Binge Drinker	Men	Women	
9	Yes	=B3/$D$5	=C3/$D$5	
10	No	=B4/$D$5	=C4/$D$5	

3. The resulting proportions are shown below. The cells can be formatted to a specific number of decimal places. The sum of the proportions should be 1. Any difference is due to a roundoff error.

7	Joint Distribution		
8	Frequent Binge Drinker	Men	Women
9	Yes	0.095	0.099
10	No	0.325	0.482

Remarks

In the **joint distribution,** the proportions of men and women frequent binge drinkers are similar. This would suggest that the number of frequent binge drinkers is similar for men and women. However, the proportion for women who are not frequent binge drinkers is higher than the proportion for men. One explanation is that there are more women in the sample than men.

Marginal Distribution

IPS Example 9.4 Marginal Distribution by Gender

1. Input another table on the same worksheet below the previous two tables.

12	Marginal Distribution of Gender		
13		Men	Women
14	Proportion		

2. Input the formulas shown below into the second table. The formulas divide the total gender count by the total sample size. This is the **marginal distribution of gender**. The reference for the total should be absolute (denoted by the $ before both the column letter and row number) if the formula is copied.

	A	B	C	D
1			Gender	
2	**Frequent Binge Drinker**	**Men**	**Women**	**Total**
3	Yes	1,630	1,684	3,314
4	No	5,550	8,232	13,782
5	Total	7,180	9,916	17,096
6				
7	Joint Distribution			
8	Frequent Binge Drinker	Men	Women	
9	Yes	=B3/$D$5	=C3/$D$5	
10	No	=B4/$D$5	=C4/$D$5	
11				
12	Marginal Distribution of Gender			
13		Men	Women	
14	Proportion	=B5/$D$5	=C5/$D$5	

3. The resulting proportions are shown below. The cells can be formatted to a specific number of decimal places or changed to a percentage. The sum of the proportions should be 1. Any difference is due to a roundoff error.

12	Marginal Distribution of Gender		
13		Men	Women
14	Proportion	0.420	0.580

Remarks

In the **marginal distribution,** the proportions by gender show that 58% of the sample was women.

IPS Example 9.5 Marginal Distribution by Frequent Binge Drinking

1. Input another table on the same worksheet below the previous three tables.

16	Marginal Distribution of Frequent Binge Drinking		
17		Yes	No
18	Proportion		

2. Input the formulas shown below into the second table. The formulas divide the total gender count by the total sample size. This is the **marginal distribution of frequent binge drinking.** In this case, the reference should not be absolute.

	A	B	C	D
1		Gender		
2	Frequent Binge Drinker	Men	Women	Total
3	Yes	1,630	1,684	3,314
4	No	5,550	8,232	13,782
5	Total	7,180	9,916	17,096
6				
7	Joint Distribution			
8	Frequent Binge Drinker	Men	Women	
9	Yes	=B3/$D$5	=C3/$D$5	
10	No	=B4/$D$5	=C4/$D$5	
11				
12	Marginal Distribution of Gender			
13		Men	Women	
14	Proportion	=B5/$D$5	=C5/$D$5	
15				
16	Marginal Distribution of Frequent Binge Drinking			
17		Yes	No	
18	Proportion	=D3/D5	=D4/D5	

3. The resulting proportions are shown below. The cells can be formatted to a specific number of decimal places or changed to a percentage. The sum of the proportions should be 1. Any difference is due to a roundoff error.

16	Marginal Distribution of Frequent Binge Drinking		
17		Yes	No
18	Proportion	0.194	0.806

Remarks

In the **marginal distribution,** the proportions by frequent binge drinker show that almost 20% (19.4%) of the sample were frequent binge drinkers.

Describing Relations in Two-Way Tables

Relationships among categorical variables are described by calculating other percents from the counts given. Which ones are calculated depends on the problem and which variable you want to focus on.

Conditional Distribution

IPS Example 9.6 Conditional Distribution for the Gender—Binge Drinking Table

The percent of women or men who are binge drinkers would be a calculation that would give insight into the relationship between these variables. This is called the **conditional distribution.**

1. Input two tables on the same worksheet below the previous four tables.

	Conditional Distribution of Frequent Binge Drinking for Women		
20			
21		Yes	No
22	Percent		
23			
24	Conditional Distribution of Frequent Binge Drinking for Men		
25		Yes	No
26	Percent		

2. Input the formulas shown below into both tables. These formulas investigate the gender distribution separately. In this case, the reference should not be absolute.

	A	B	C	D
1		Gender		
2	Frequent Binge Drinker	Men	Women	Total
3	Yes	1,630	1,684	3,314
4	No	5,550	8,232	13,782
5	Total	7,180	9,916	17,096
7	Joint Distribution			
8	Frequent Binge Drinker	Men	Women	
9	Yes	=B3/$D$5	=C3/$D$5	
10	No	=B4/$D$5	=C4/$D$5	
11				
12	Marginal Distribution of Gender			
13		Men	Women	
14	Proportion	=B5/$D$5	=C5/$D$5	
15				
16	Marginal Distribution of Frequent Binge Drinking			
17		Yes	No	
18	Proportion	=D3/D5	=D4/D5	
19				
20	Conditional Distribution of Frequent Binge Drinking for Women			
21		Yes	No	
22	Percent	=C3/C5	=C4/C5	
23				
24	Conditional Distribution of Frequent Binge Drinking for Men			
25		Yes	No	
26	Percent	=B3/B5	=B4/B5	

3. The resulting proportions are shown below. The cells can be formatted to a specific number of decimal places or changed to a percentage.

	Conditional Distribution of Frequent Binge Drinking for Women		
20			
21		Yes	No
22	Percent	17.0%	83.0%
23			
24	Conditional Distribution of Frequent Binge Drinking for Men		
25		Yes	No
26	Percent	22.7%	77.3%

Remarks

A comparison of the **conditional distributions** gives us insight into the relationship between gender and frequent binge drinking. For this sample, the men are more likely to be frequent binge drinkers than the women.

9.2   Inference for Two-Way Tables

A statistical test is used to determine if results would normally occur by chance or if they are significant, that is, outside of an accepted limit determined by a confidence level, such as 95%. We need to use an appropriate statistical test to determine if the results in the conditional distribution between men and women can be attributed to chance alone.

The null hypothesis (H_0) for a two-way table is that there is no relationship between the row variable and the column variable. The alternative hypothesis (H_a) is that there is an association between the variables; however, it does not specify a direction and is not described as either one-sided or two-sided.

In the previous example, the null hypothesis would be that there is no association between the gender and whether or not a person is a frequent binge drinker. The alternative hypothesis would be that there is an association between the two variables.

IPS Examples 9.1 & 9.14   Exclusive Territories and Franchise Firm Success

Many popular businesses are franchises. The relationship between the local entrepreneur and the franchise firm is spelled out in a detailed contract. One clause that the contract may or may not contain is the entrepreneur's right to an exclusive territory. This means that the new outlet will be the only representative of the franchise in a specified territory and will not have to compete with other outlets of the same chain. How does the presence of an exclusive territory clause in the contract relate to the survival of the business? A study collected data from a sample of 170 new franchise firms.

Two categorical variables were measured for each firm. First, the firm was classified as successful or not based on whether or not it was still franchising as of a certain date. Second, the contract each firm offered to franchisees was classified according to whether or not there was an exclusive territory clause. The explanatory variable is the Exclusive Territory clause and is the column variable. The row variable is the response, in this case Success. Here are the data:

	A	B	C	D
1		Exclusive Territory		
2	Success	Yes	No	Total
3	Yes	108	15	123
4	No	34	13	47
5	Total	142	28	170

1. Input the table as shown above into an empty worksheet.

2. Examine column percents by inputting the following formulas into a second table created as shown below.

	A	B	C	D
1		Exclusive T		
2	Success	Yes	No	Total
3	Yes	108	15	123
4	No	34	13	47
5	Total	142	28	170
6				
7	Column Pe			
8		Exclusive T		
9	Success	Yes	No	Total
10	Yes	=B3/$B$5	=C3/$C$5	=D3/$D$5
11	No	=B4/$B$5	=C4/$C$5	=D4/$D$5
12	Total	=B5/$B$5	=C5/$C$5	=D5/$D$5

3. The resulting percents are shown below. The cells can be formatted to a specific number of decimal places.

	A	B	C	D
1		Exclusive Territory		
2	Success	Yes	No	Total
3	Yes	108	15	123
4	No	34	13	47
5	Total	142	28	170
6				
7	Column Percents for Firms			
8		Exclusive Territory		
9	Success	Yes	No	Total
10	Yes	76.1%	53.6%	72.4%
11	No	23.9%	46.4%	27.6%
12	Total	100.0%	100.0%	100.0%

Remarks

There is a large difference between the percents of successes in both groups. A statistical test will tell us whether this difference is unlikely to have occurred by chance.

Expected Cell Counts

To test the null hypothesis for two-way tables, the *observed* values or counts in the table are compared to *expected* values or counts. The test statistic used for two-way tables is a numerical measure of the difference between the observed and the expected cell counts.

The expected count for any cell of a two-way table is defined as:

$$\text{expected count} = \frac{\text{row total} \times \text{column total}}{n}$$

where n is the total number of observations.

The Chi-Square Test

The association between two variables in a two-way table is determined by calculating a statistic that compares the entire set of observed counts with the entire set of calculated expected counts. The result is called the **chi-squared statistic, X^2** (from the Greek letter chi).

$$X^2 = \sum \frac{(\text{observed count} - \text{expected count})^2}{\text{expected count}}$$

If there is a large difference between the observed and expected values, a large X^2 will result. The probability resulting from this statistic is called the chi-square distribution. As with the t distribution, there are a series of X^2 distributions determined by the degrees of freedom. The X^2 test always uses the upper tail of the X^2 distribution because any deviation from the null hypothesis makes the statistic larger.

This approximation is used for tables larger than 2×2 with all expected cell counts of at least five or more.

IPS Example 9.16 (Using Example 9.1 data) Exclusive Territories and Franchise Firm Success

The chi-squared test can be conducted on the existing data set for the exercise used previously. The result of the test will indicate whether an exclusive territories contract has an association with a franchise firm's success.

1. Input the following formulas to calculate the row and column percentages for the data table.

	A	B	C	D	E
1		Exclusive Ter			
2	Success	Yes	No	Total	
3	Yes	108	15	123	
4	No	34	13	47	
5	Total	142	28	170	
6					
7	Column Per				
8		Exclusive Ter			
9	Success	Yes	No	Total	
10	Yes	=B3/$B$5	=C3/$C$5	=D3/$D$5	
11	No	=B4/$B$5	=C4/$C$5	=D4/$D$5	
12	Total	=B5/$B$5	=C5/$C$5	=D5/$D$5	
13					
14		Exclusive Ter			
15	Success	Yes	No	Total	Cell format:
16	Yes	108	15	123	Count
17		=B16/$D$16	=C16/$D$16	=D16/$D$16	Row Percent
18		=B16/B24	=C16/C24	=D16/D24	Column Percent
19		=B16/$D$24	=C16/$D$24	=D16/$D$24	Total Percent
20	No	34	13	47	
21		=B20/$D$20	=C20/$D$20	=D20/$D$20	
22		=B20/B24	=C20/C24	=D20/D24	
23		=B20/$D$24	=C20/$D$24	=D20/$D$24	
24	Total	142	28	170	
25		=B24/$D$24	=C24/$D$24	=D24/$D$24	
26		=B24/B24	=C24/C24	=D24/D24	
27		=B24/$D$24	=C24/$D$24	=D24/$D$24	

2. The resulting percents are shown below. The cells can be formatted to a specific number of decimal places or changed to a percentage.

		Exclusive Territory			
14					
15	Success	Yes	No	Total	Cell format:
16	Yes	108	15	123	Count
17		87.8%	12.2%	100.0%	Row Percent
18		76.1%	53.6%	72.4%	Column Percent
19		63.5%	8.8%	72.4%	Total Percent
20	No	34	13	47	
21		72.3%	27.7%	100.0%	
22		23.9%	46.4%	27.6%	
23		20.0%	7.6%	27.6%	
24	Total	142	28	170	
25		83.5%	16.5%	100.0%	
26		100.0%	100.0%	100.0%	
27		83.5%	16.5%	100.0%	

3. In order to calculate the chi-squared test, the expected values need to be calculated. For the success of the franchise-exclusive contract table, we multiply the proportion of successful firms (123/170) by the number of firms with exclusive territories (142). This result is 102.74. Based on the proportions, this is the expected value for the first cell, intersecting the first column and the first row. The actual value is 108. Complete the rest of the formulas as follows.

	A	B	C	D	E	F	G	H	I
1		Exclu:					Expected Counts		
2	Success	Yes	No	Total		Success	Yes	No	Total
3	Yes	108	15	123		Yes	=$B$5*$D$3/$D$5	=$C$5*$D$3/$D$5	=SUM(G3:H3)
4	No	34	13	47		No	=$B$5*$D$4/$D$5	=$C$5*$D$4/$D$5	47
5	Total	142	28	170		Total	142	28	170

4. The resulting values are shown below. The cells can be formatted to a specific number of decimal places.

F	G	H	I
	Expected Counts		
Success	Yes	No	Total
Yes	102.74	20.26	123.00
No	39.26	7.74	47.00
Total	142.00	28.00	170.00

5. Calculate the chi-squared test by inputting the **CHITEST** function as shown below. The **CHITEST** function compares the actual counts to the calculated expected counts.

F	G	H	I
	Expected Counts		
Success	Yes	No	Total
Yes	102.74	20.26	123.00
No	39.26	7.74	47.00
Total	142.00	28.00	170.00
Chi-Square Test			
=CHITEST(B3:C4,G3:H4)			

6. The resulting *P*-value is shown below. The result can be formatted to a specific number of decimal places.

Chi-Square Test
0.01505

Remarks

The chi-squared test yields a *P*-value of 0.015 or 1.5%, which is statistically significant at the 5% level. This provides evidence for the alternative hypothesis, which states that there is a relationship between the success of a franchise and whether or not the franchise has exclusive territory rights. We cannot tell whether having an exclusive territories contract is the cause of success of a franchise firm. There may be other confounding variables contributing to this relationship.

9.3   Formulas and Models for Two-Way Tables

This section outlines the calculations used to analyze a two-way table. The steps taken to analyze the table in Section 9.1 are summarized below:

1. Calculate row and/or column percents to assist in describing the data set.
2. Calculate the expected counts and the associated X^2 statistic.
3. Find the associated *P*-value from the calculated statistic.
4. Draw a conclusion about the association between the row and column variables.

IPS Example 9.18 Background Music and Consumer Behavior

Market researchers know that background music can influence the mood and purchasing behavior of customers. One study in a supermarket in Northern Ireland compared three treatments: no music, French accordion music, and Italian string music. Under each condition, the researchers recorded the numbers of bottles of French, Italian, and other wine purchased. The summarized data is shown below:

Wine	None	Music French	Italian	Total
French	30	39	30	99
Italian	11	1	19	31
Other	43	35	35	113
Total	84	75	84	243

This data set is a 3 × 3 table with marginal totals. It is suspected that the type of background music can explain the sale of different types of wine. The conditional distribution will help investigate this association.

Two-Way Table Calculation—Column Percentages

1. Input the data table into Excel including **Totals.**

2. Calculate the column percentages by inputting a second table and formulas as shown below.

	A	B	C	D	E
1			Music		
2	Wine	None	French	Italian	Total
3	French	30	39	30	99
4	Italian	11	1	19	31
5	Other	43	35	35	113
6	Total	84	75	84	243
7					
8			Music		
9	Wine	None	French	Italian	Total
10	French	=B3/B6	=C3/C6	=D3/D6	=E3/E6
11	Italian	=B4/B6	=C4/C6	=D4/D6	=E4/E6
12	Other	=B5/B6	=C5/C6	=D5/D6	=E5/E6
13	Total	=B6/B6	=C6/C6	=D6/D6	=E6/E6

3. The resulting values are shown below. The cells can be formatted to a specific number of decimal places.

7	Column Percentages				
8			Music		
9	Wine	None	French	Italian	Total
10	French	35.7%	52.0%	35.7%	40.7%
11	Italian	13.1%	1.3%	22.6%	12.8%
12	Other	51.2%	46.7%	41.7%	46.5%
13	Total	100.0%	100.0%	100.0%	100.0%

Remarks

The results show that 35.7% of the wine sold was French when no music was played, while 52% of the wine sold was French when French music was played. All columns add up to 100%.

Graphing Percentages

1. Select the three column percentages (35.7%, 13.1%, 51.2%) in the first column (No Music) by selecting 35.7%, holding the Ctrl key down and selecting the other two percentages.

2. Click the **ChartWizard** in the toolbar.

3. Select **Column Graph** and click **Next.**

4. **Step 2** of the **ChartWizard** will display the three percentages graphed. To add their labels, click the **Series** tab at the top of the dialog box.

5. Click inside the box for **Category (X) axis labels** and click the **Collapse Dialog** box icon to select the labels on your original worksheet.

6. Click back on the worksheet to return to your original data sheet and select the three wine labels.

Wine	None	French	Italian	Total
French	30	39	30	99
Italian	11	1	19	31
Other	43	35	35	113
Total	84	75	84	243

7. Click the **Collapse Dialog** box icon again to return to the **Series** dialog box.

8. The appropriate labels should now appear as the *x* axis labels on the preview chart. Click **Next.**

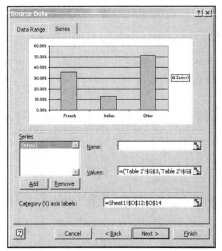

9. Add a title and appropriate axes labels.

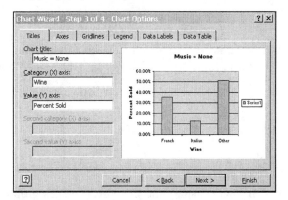

10. To de-select the **Legend**, select the **Legend** tab and de-select **Show Legend.** Click **Next**. Select the chart to be placed **As Object in** the existing table of results. Click **Finish.**

11. Continue the process to create three column graphs depicting the resulting column percentages. The format for the percentages can be changed by double-clicking on the *y* axis percentages and selecting **0** for **Decimal Places**. Click **OK.**

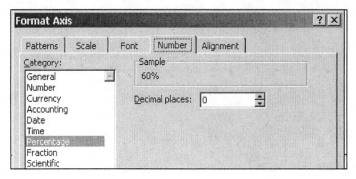

12. The gridlines can be cleared by right-clicking on the gridlines and selecting **Clear.**

13. To compare the three charts to each other, the same *y* scale should be used. If any of the scales need to be adjusted, double-click on the *y* axis and change the maximum values. Click **OK.**

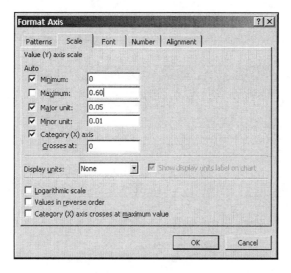

Remarks

There appears to be some relationship between the type of music played and the number and type of bottles of wine sold. When French music is played, the sale of Italian wine is very low but when Italian music is played, the sale of Italian wine increases. French wine sold well under all conditions.

Examining the corresponding row percentages can enhance the analysis of this dataset.

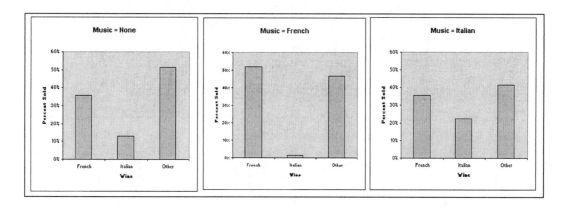

Two-Way Table Calculation—Row Percentages

IPS Example 9.18 Continued

1. Input another data table into Excel including **Totals.**

2. Calculate the row percentages by inputting a second table and formulas as shown below.

	A	B	C	D	E
1			Music		
2	Wine	None	French	Italian	Total
3	French	30	39	30	99
4	Italian	11	1	19	31
5	Other	43	35	35	113
6	Total	84	75	84	243
7	Column Pe				
8			Music		
9	Wine	None	French	Italian	Total
10	French	=B3/B6	=C3/C6	=D3/D6	=E3/E6
11	Italian	=B4/B6	=C4/C6	=D4/D6	=E4/E6
12	Other	=B5/B6	=C5/C6	=D5/D6	=E5/E6
13	Total	=B6/B6	=C6/C6	=D6/D6	=E6/E6
14	Row Perce				
15			Music		
16	Wine	None	French	Italian	Total
17	French	=B3/E3	=C3/E3	=D3/E3	=E3/E3
18	Italian	=B4/E4	=C4/E4	=D4/E4	=E4/E4
19	Other	=B5/E5	=C5/E5	=D5/E5	=E5/E5
20	Total	=B6/E6	=C6/E6	=D6/E6	=E6/E6

3. The resulting values are shown below. The cells can be formatted to a specific number of decimal places.

14	Row Percentages				
15			Music		
16	Wine	None	French	Italian	Total
17	French	30.3%	39.4%	30.3%	100.0%
18	Italian	35.5%	3.2%	61.3%	100.0%
19	Other	38.1%	31.0%	31.0%	100.0%
20	Total	34.6%	30.9%	34.6%	100.0%

4. Create the same column graphs in Excel using the previous method.

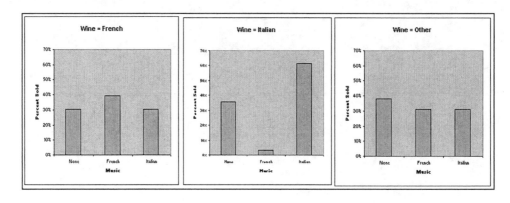

Remarks

The row percentage comparison shows that more French wine is purchased while French music is played, and more Italian wine is purchased when Italian music is played. There is a strong negative effect of French music on the sale of Italian wine.

Chi-Squared (X^2) Calculation

IPS Example 9.18 Continued

1. Calculate the expected values using the same method as the previous example. The chi-squared test results give a *P*-value of 0.

Actual						1	2	3	Totals		Rows:	Type of Wine
30	39	30	99	Counts		30	39	30	99	1	Columns:	Type of Music
11	1	19	31	Exp Freq		34.22	30.56	34.22	99.00			
43	35	35	113									
84	75	84	243									
						11	1	19	31	2		
Expected						10.72	9.57	10.72	31.00			
34.22	30.56	34.22	99.00									
10.72	9.57	10.72	31.00									
39.06	34.88	39.06	113.00			43	35	35	113	3		
84.00	75.00	84.00	243.00			39.06	34.88	39.06	113.00			
Chi-Squared	P-Value	DF										
	0.001088	4				84	75	84	243	Totals		
						84.00	75.00	84.00	243.00			

Remarks

The resulting low *P*-value indicates that the results are significant and that there is a strong relationship between the type of background music played and the number of bottles of specific types of wine sold. In examining the expected frequencies, we can see two values that contribute heavily to the calculation of the X^2 statistic—the sales of Italian wine are much lower than expected when French music is played and much above what is expected when Italian music is played. One specific conclusion is that the sale of Italian wine is strongly affected by Italian and French music.

Chapter 10

Inference for Regression

Predicting a response from one or several explanatory variables is a common statistical procedure. This prediction is straightforward when both variables are quantitative and the data follow a straight line. This method was discussed previously in Chapter 2 and will be expanded upon in this chapter.

Simple linear regression is defined as the straight-line relationship between a quantitative response variable and a single quantitative explanatory variable. Regression involving more than one quantitative variable is called **multiple regression** and will be addressed in Chapter 11.

When a linear relationship exists between the explanatory x variable and the response y variable, we use the least-squares regression line to fit the data and make the predictions. In this chapter, we will conduct significance tests and calculate confidence intervals for the regression line. The least-squares equation constructed from sample data is an estimation of the regression line of the population. This is consistent with our previous analysis estimating population parameters, such as the population mean μ from sample statistics such as $\bar{x}$.

10.1   Simple Linear Regression

The population regression line can be written as $\beta_0 + \beta_1 x$. β_0 and β_1 are parameters that describe the straight-line relationship, with β_0 as the y-intercept and β_1 as the slope. The sample regression line can be written as $b_0 + b_1 x$ with the sample statistics b_0 as the y-intercept and b_1 as the slope. We can calculate confidence intervals and significance tests for inference about the slope β_1 and the y-intercept β_0. Each predicted response for the estimation of the population parameter has a variation about the predicted value that can be represented by a normal distribution with the same standard deviation. The **true population regression line** is represented by the means of those predicted values:

$$\mu_y = \beta_0 + \beta_1 x$$

IPS Examples 10.1 & 10.2 Does Speed Affect Fuel Efficiency?

Computers in some vehicles calculate various quantities related to the performance. One of these is the fuel efficiency, or gas mileage, expressed as miles per gallon (MPG). Another is the average speed in miles per hour (MPH). For one vehicle equipped in this way, MPG and MPH were recorded each time the gas tank was filled and the computer was reset. How does the speed at which the vehicle is driven affect the fuel efficiency? There are 234 observations available. We will work with a simple random sample of size 60.

Creating a Scatterplot and Predicted Values

1.   Open the file **eg10_001** from the **IPS CD-ROM.**

2.   Create an XY scatterplot with MPH values as the explanatory variable and MPG as the response variable. The data will have to be copy and pasted in a new location of the worksheet and in the correct XY order.

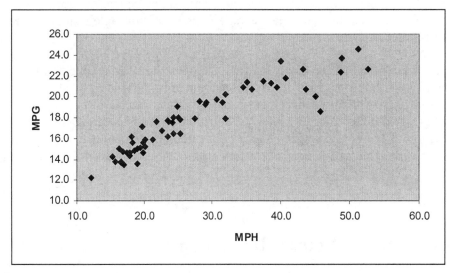

Remarks

The scatterplot shows a fairly linear relationship up through 30 MPH, but beyond that point, the data start to curve and are more scattered. It is decided to apply a transformation to approximate a linear relationship.

IPS Example 10.3 Transform Data to Approximate a Linear Relationship

One type of function that will apply a smooth function fit is a logarithm. Therefore, we will examine the effect of transforming speed by taking the natural logarithm of the MPH values.

Applying a Logarithmic Transformation

1. Using the file **eg10_001** data from the previous example, insert a column to the right of MPH entitled LN MPH. Insert the LN function of the MPH data as shown below.

MPH	LN MPH
18.6	=LN(G2)
19.5	=LN(G3)
24.2	=LN(G4)
17.9	=LN(G5)
21.2	=LN(G6)
32	=LN(G7)
17	=LN(G8)

2. Create an XY Scatterplot with LN MPH on the *x* axis and MPG on the *y* axis.

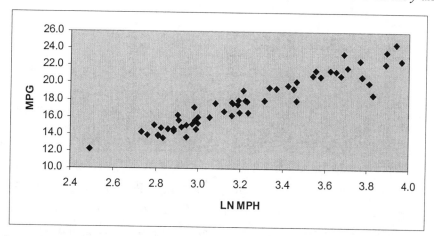

3. Insert a trendline (least-squares regression) with linear formula by right-clicking on any data point and select **Add Trendline.** Accept the default linear **Type.** Select the **Options** tab and select **Display equation on chart.** Click **OK.**

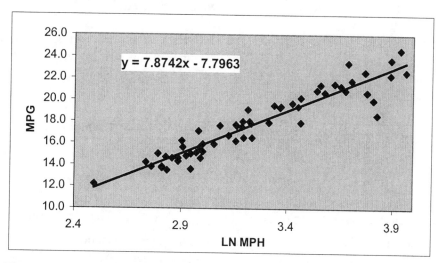

4. The predicted regression line equation can be used to predict the MPG (*y*) for any given MPH (*x*) by using the following method:

- Create a column to the right of the MPG column, titled **Predicted MPG.**

- Insert the trendline formula into the first cell for the **Predicted MPG** column as shown below.

G	H	I	J
MPH	**LN MPH**	**MPG**	**Predicted MPG**
18.6	2.9	14.8	=7.8742*H2-7.7963

- Copy the formula down to the bottom of the column.

- Note: residuals can be automatically calculated using the **Regression Tool** under **Tools/Data Analysis.**

Simple Linear Regression Model

The sample statistic least-squares line would take somewhat different values for slope and y-intercept if the study was repeated. In formal inference, we say that the sample slope and y-intercept are estimates of the unknown parameters. For any fixed explanatory x-value, the response y-value varies according to a normal distribution. The population regression line representing the mean response, $\mu_y = \beta_0 + \beta_1 x$, relates the MPG μ_y with MPH x. The slope β_1 represents the mean increase in MPG for each MPH. The y-intercept β_0 represents the values at $x = 0$. Both β_1 and β_0 are unknown parameters. The standard deviation of the response y-value (σ) is the same for all values of x and is also an unknown parameter. The population regression line describes the "on-the–average" relationship between x and y. The values of y that we can observe vary about their means as a normal distribution. The standard deviation determines whether the points fall close to the true regression line or are widely scattered.

The sample least-squares line estimates the true regression line and its calculated residuals represent the vertical deviations from the line. In this model, we use ε (Greek epsilon) to represent the residual. A response value y is the sum of its mean value and the vertical chance deviation ε from the mean.

The statistical model for simple linear regression states that the observed response y can be predicted from the explanatory variable x by the relationship

$$y_i = \beta_0 + \beta_1 x_i + \varepsilon_i$$

where the mean response is represented by $\mu_y = \beta_0 + \beta_1 x$. The deviations ε_i are independent and normally distributed with mean 0 and standard deviation σ.

Estimating the Regression Parameters

The least-squares line $\hat{y} = b_0 + b_1 x$ estimates the slope β_1 and the intercept β_0 of the population regression line. The slope of the least-squares line is represented by

$$b_1 = r\frac{s_y}{s_x}$$

and the intercept by $b_0 = \bar{y} - b_1\bar{x}$, where r is the correlation between the observed values of y and x, s_y is the standard deviation of the sample of y's, and s_x is the standard deviation of the sample of x's.

The standard deviation σ measures the variation of y about the population regression line. We will use the deviations of the observed responses from the least-squares line to estimate σ.

We will use e_i as the residual for the ith observation:

$$e_i = \text{observed response} - \text{predicted response}$$
$$= y_i - \hat{y}_i$$
$$= y_i - b_0 - b_1 x_i$$

where e_i are the sample values that correspond to the model values ε_i and sum to 0.

For simple linear regression, the estimate of σ^2 is the average squared residual:

$$s^2 = \frac{1}{n-2}\sum e_i^2$$

$$s^2 = \frac{1}{n-2}\sum(y_i - \hat{y}_i)^2$$

The residuals and s^2 have $n - 2$ degrees of freedom. The standard deviation σ is estimated by $s = \sqrt{s^2}$, which is called the regression standard error.

Therefore, the standard deviation σ in the statistical model is estimated by the regression standard error:

$$s^2 = \sqrt{\frac{1}{n-2}\sum(y_i - \hat{y}_i)^2}$$

Estimating Regression Parameters
Plotting Residuals

IPS Example 10.4 Relationship of MPG and MPH

We will now investigate this data set using the Excel analysis tool.

1. Select **Tools** ⇨ **Data Analysis** from the menu and **Regression** from the list of **Analysis Tools**. Click **OK**.

2. Select the *x* and *y* ranges including labels, select the **Labels** box, the **Confidence Interval** box, **New Worksheet** for the **Output** option, and **Residuals, Residual Plots**, and **Line Fit Plots**. Click **OK**.

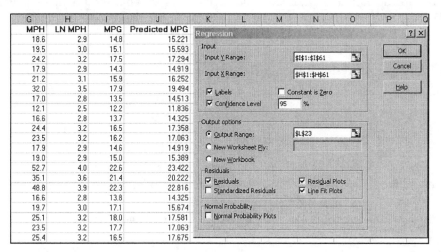

3. The results are shown on a new worksheet and, after adjusting the column widths to display the information, can be analyzed.

SUMMARY OUTPUT

Regression Statistics	
Multiple R	0.946053
R Square	0.8950163
Adjusted R Square	0.8932062
Standard Error	0.9995164
Observations	60

ANOVA

	df	SS	MS	F	Significance F
Regression	1	493.9885883	493.9886	494.4668	4.50949E-30
Residual	58	57.94391174	0.999033		
Total	59	551.9325			

	Coefficients	Standard Error	t Stat	P-value	Lower 95%	Upper 95%	Lower 95.0%	Upper 95.0%
Intercept	-7.7962501	1.154944262	-6.75033	7.69E-09	-10.10812052	-5.4843797	-10.108121	-5.48437974
LN MPH	7.874219	0.354110611	22.23661	4.51E-30	7.165390143	8.58304788	7.16539014	8.583047883

4. The coefficients are identified as Intercept (b_0) = -7.796, the slope as 7.874, and the standard error as 1.155.

5. The axes of the residual plots can be re-scaled to examine any patterns.

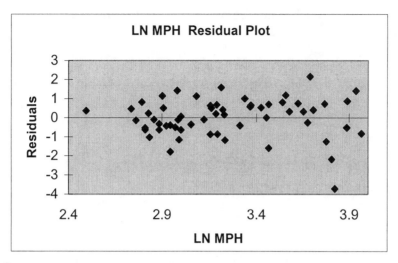

Remarks

No clear pattern is evident in this residual plot. One residual is low and corresponds to the outlier already identified.

Confidence Intervals and Significance Tests

Confidence intervals and tests for the slope and intercept are based on the sampling distributions of the estimates b_1 and b_0. When the simple linear regression model is true, each of b_1 and b_0 has a normal distribution. The mean of b_0 is β_0 and the mean of b_1 is β_1, which means that the intercept and the slope of the fitted line are unbiased estimates of the intercept and slope of the population regression line. Regression inference is robust against even a moderate lack of normality in the data. Outliers and other influential observations can, however, invalidate the results.

A level C confidence interval for the slope β_1 of the population regression line equals

$$b_1 \pm t * SE_{b_1}$$

where t^* is the critical t distribution value for the degrees of freedom of $(n-2)$ with area C between $-t^*$ and t^*.

The test statistic is $t = \dfrac{b_1}{SE_{b_1}}$ where the null hypothesis equals $\mu_y = \beta_0$, which says that the mean of y does not vary with x. This is the result of saying that $\beta_1 = 0$ and there is no straight-line relationship between x and y and no value to predicting y. The t test for this hypothesis investigates whether the regression relationship between x and y is large enough to be statistically significant.

IPS Examples 10.5 & 10.6 Relationship Between Fuel Economy and Speed Continued

The computer output for Example 10.4 contains the information needed for inference about the regression slope and intercept.

The t statistic is calculated as 22.82, yielding a P-value of nearly zero. In addition, the 95% confidence interval for β_1 is shown as (7.16, 8.58).

Remarks

The results are statistically significant at the 1% level, which indicates that there is a strong relationship between speed and fuel efficiency.

Prediction Intervals

In previous sections, we determined that it was straightforward to calculate a prediction from the least-squares regression line. The method is to simply substitute the desired x-value into the equation and solve for the predicted y-value.

In the previous example for weekly earnings of female bank workers for length of service, the least-squares equation is

$$\hat{y} = 349.4 + 0.5905x$$

To predict earnings for a worker with 125 months of service, we substitute 125 for the x-value and solve.

We can define an interval that describes the accuracy of this prediction. However, the prediction yields two different margins of error if you are referring to all the workers with 125 months of service or one worker with 125 months of service. Individual workers with 125 months of service don't all earn the same wages, so there needs to be a larger margin of error than the estimate for all workers with 125 months of service.
The given value of the explanatory variable x is defined as x^*. In the above example, $x^* = 125$. The difference between the prediction for a single outcome and the prediction for the mean of all outcomes when $x = x^*$ is the margin of error.

To estimate the mean response, we use a confidence interval of the form:

$$\hat{y} \pm t^* SE_{\hat{\mu}}$$

where $SE_{\hat{\mu}}$ is the standard error for estimating a mean response:

$$SE_{\hat{\mu}} = s \sqrt{\frac{1}{n} + \frac{(x^* - \bar{x})^2}{\sum(x - \bar{x})^2}}$$

where the sum includes all observations for the explanatory variable x.

A level C prediction interval for a single observation on y when x has the value x^* is

$$\hat{y} \pm t^* SE_{\hat{y}}$$

where $SE_{\hat{y}}$ is the standard error for predicting an individual response:

$$SE_{\hat{y}} = s \sqrt{1 + \frac{1}{n} + \frac{(x^* - \bar{x})^2}{\sum(x - \bar{x})^2}}$$

The prediction interval is wider than the confidence interval because the standard error is larger. For both methods, t^* is the critical value for $(n - 2)$ degrees of freedom with area C between $-t^*$ and t^*.

IPS Example 10.10 Predicting Future Observation of Fuel Efficiency

Using our previous fuel efficiency data, we will find the prediction interval for a future observation of fuel efficiency when the vehicle is driven at 30 MPH.

1. Any MPG predicted value can be calculated using the regression formula

   $$MPG = -7.80 + (7.87)(3.4) = 19.0$$

This value is within the 95% prediction interval calculated to be between 17.0 MPG and 21.0 MPG.

Chapter 11

Multiple Regression

Thus far, we have examined the relationship for a single explanatory variable and a single response variable. It is often the case that a response is explained by more than one factor or explanatory variable. In developing the techniques for multiple regression analysis, we will utilize many of the methods and procedures developed in some of the previous chapters.

11.1   Inference for Multiple Regression

In multiple regression, the response variable y depends on more than one (up to p) explanatory variables denoted by $x_1, x_2, ..., x_p$. The mean of the response variable y is a linear function of the explanatory variables.

$$\mu_y = \beta_0 + \beta_1 x_1 + \beta_2 x_2 + ... + \beta_p x_p$$

This equation is referred to as the **population regression equation**.

IPS Example 11.1   Data for Student Test Performance

Our case study uses data collected at a large university on all first-year computer science majors in a particular year. The purpose of the study was to attempt to predict success in the early university years. One measure of success was the cumulative grade point average (GPA) after three semesters. Among the explanatory variables recorded at the time the students enrolled in the university were average high school grades in mathematics (HSM), science (HSS), and English (HSE). We will use high school grades to predict the response variable GPA. There are $p = 3$ explanatory variables: $x_1 = $ HSM, $x_2 = $ HSS, and $x_3 = $ HSE. The high school grades are coded on a scale from 1 to 10, with 10 corresponding to A, 9 to A−, 8 to B+, and so on. These grades define the subpopulations.

One possible multiple regression model for the subpopulation mean GPAs is

$$\mu_{GPA} = \beta_0 + \beta_1 HSM + \beta_2 HSS + \beta_3 HSE$$

 The following table defines the notation, which identifies the first subscript as the individual and the second as the explanatory variable. The full data set is in the CSDATA data set in the Data Appendix of the CD-ROM.

	Variables				
Individual	x_1	x_2	....	x_p	y
Student 1	x_{11}	x_{12}	....	x_{1p}	y_1
Student 2	x_{21}	x_{22}	....	x_{2p}	y_2
....			....		
Student n	x_{n1}	x_{n2}	....	x_{np}	y_n

Where n is the number of individuals or observations ($n = 234$ in this data set) and p is the number of explanatory variables ($p = 3$ in this data set).

Multiple Linear Regression Model

As in Chapter 10, the sample least-squares line estimates the true regression line and its calculated residuals (ε) represent the vertical deviations from the line. A response value y is the sum of its mean value and a chance deviation ε from the mean. Otherwise phrased as DATA = FIT + RESIDUAL. That is,

$$y = \mu_y + \varepsilon$$

The deviations ε are independent and Normally distributed with mean 0 and standard deviation σ.

Expanding the statistical model for simple linear regression to address multiple explanatory variables, the statistical model for multiple linear regression is for $i = 1, 2,..., n$:

$$y_i = \beta_0 + \beta_1 x_{i1} + \beta_2 x_{i2} + ... + \beta_p x_{ip} + \varepsilon_i$$

where the mean response is represented by $\mu_y = \beta_0 + \beta_1 x_1 + \beta_2 x_2 + ... + \beta_p x_p$.

A residual is still defined as the difference between the observed and predicted response. The deviations ε_i are independent and are an SRS from the $N(0, \sigma)$ distribution. For the ith residual, the formula is:

$$\varepsilon_i = \text{observed response} - \text{predicted response}$$
$$= y_i - \hat{y}_i$$
$$= y_i - b_0 - b_1 x_{i1} - b_2 x_{i2} - ... - b_p x_{ip}$$

The least-squares regression method minimizes the sum of the squares of the residuals. In multiple regression, this is the same as saying that the coefficients $b_0, b_1, b_2, ..., b_p$ are values that minimize the resulting residuals.

The population parameters that are estimated from the sample data are $\beta_0, \beta_1, ..., \beta_p$ and σ. The parameters are estimated by the coefficients in the multiple regression equation:

$$\hat{y} = b_0 + b_1 x_{i1} + b_2 x_{i2} + ... + b_p x_{ip}$$

The variance is defined by $s^2 = \dfrac{\sum e_i^2}{n - p - 1}$ and the regression standard error is $s = \sqrt{s^2}$.

where $(n - p - 1)$ are the **degrees of freedom** for s^2, where p is the number of explanatory variables.

Confidence Intervals and Significance Tests for Regression Coefficients

A level C confidence interval for β_i is $b_j \pm t^* SE_{bj}$ where SE_{bj} is the standard error of b_j and t^* is the critical t-value for $(n - p - 1)$ degrees of freedom with area C between $- t^*$ and t^*.

To test the hypothesis H$_0$: $\beta_j = 0$, we compute the t statistic:

$$t = \frac{b_j}{SE_{bj}}$$

Since regression is often used for prediction, we could construct confidence intervals for a mean response and prediction intervals for future observations obtained from multiple regression models.

11.2   A Case Study

Preliminary Analysis

Multiple regression analysis is illustrated in this section by analyzing the data from Example 11.1. The response variable is the cumulative GPA after three semesters for a group of computer science majors at a large university. The explanatory variables are average high school grades, represented by HSM, HSS, and HSE. SAT mathematics and verbal scores are also examines as explanatory variables. There are $n = 224$ students in the study.

A first step in the analysis is to examine the basic statistics for each of the variables.

Analyzing Descriptive Statistics

1. Open the file **csdata** from the Appendix data folder on the **IPS CD-ROM.**

2. Select **Tools** ⇨ **Data Analysis** from the menu and **Descriptive Statistics** from the list of **Analysis Tools.** Click **OK.**

3. Select the data range for GPA, HSM, HSS, HSE, SATM, and the SATV columns, including the labels. Select the **Labels** box, a **New Worksheet** for **Output,** and **Summary Statistics.** Click **OK.**

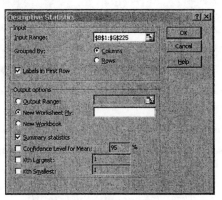

4. Adjust the column widths to see the data clearly.

Remarks

The mean SATM score is higher than the mean SATV score, which is unusual unless the group were scientists or engineers. The two standard deviations are about the same. The means of the three high school grade variables are similar with mathematics being a little higher. The mean GPA is 2.635 on a 4-point scale, with standard deviation of 0.779.

gpa		hsm		hss		hse		satm		satv	
Mean	2.63522	Mean	8.32143	Mean	8.08929	Mean	8.09375	Mean	595.286	Mean	504.549
Standard Error	0.05208	Standard Error	0.10949	Standard Error	0.11356	Standard Error	0.10075	Standard Error	5.77294	Standard Error	6.1878
Median	2.74	Median	9	Median	8	Median	8	Median	600	Median	490
Mode	3.06	Mode	10	Mode	10	Mode	9	Mode	640	Mode	480
Standard Deviation	0.77939	Standard Deviation	1.63874	Standard Deviation	1.69966	Standard Deviation	1.50787	Standard Deviation	86.4014	Standard Deviation	92.6105
Sample Variance	0.60746	Sample Variance	2.68546	Sample Variance	2.88885	Sample Variance	2.27368	Sample Variance	7465.21	Sample Variance	8576.7
Kurtosis	0.36482	Kurtosis	0.67309	Kurtosis	-0.39203	Kurtosis	0.00131	Kurtosis	0.03174	Kurtosis	0.02317
Skewness	-0.6895	Skewness	-0.9962	Skewness	-0.63835	Skewness	-0.60417	Skewness	-0.17901	Skewness	0.25426
Range	3.88	Range	8	Range	7	Range	7	Range	500	Range	475
Minimum	0.12	Minimum	2	Minimum	3	Minimum	3	Minimum	300	Minimum	285
Maximum	4	Maximum	10	Maximum	10	Maximum	10	Maximum	800	Maximum	760
Sum	590.29	Sum	1864	Sum	1812	Sum	1813	Sum	133344	Sum	113019
Count	224	Count	224	Count	224	Count	224	Count	224	Count	224

With three explanatory variables, we also have three pairs of variables to examine. We can examine the correlations and the corresponding scatterplots.

Analyzing Relationships Between Two Variables

1. Open the file **csdata** file used in the previous example.

2. Select **Tools** ⇨ **Data Analysis** from the menu and **Correlation** from the list of **Analysis Tools.** Click **OK.**

3. Select the data range for the GPA, HSM, HSS, HSE, SATM, and SATV columns including the labels, select the **Labels** box, and the **Output** on a **New Worksheet.** Click **OK.**

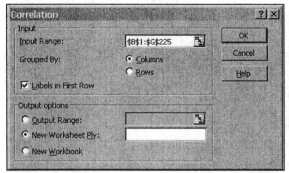

4. The results are shown with the variables listed on both the row labels and column labels. The correlations are read between the two variables intersecting.

	A	B	C	D	E	F	G
1		gpa	hsm	hss	hse	satm	satv
2	gpa	1					
3	hsm	0.436499	1				
4	hss	0.329425	0.575686	1			
5	hse	0.289001	0.446887	0.579375	1		
6	satm	0.251714	0.453514	0.240479	0.108285	1	
7	satv	0.11449	0.22112	0.261698	0.243715	0.463942	1

Remarks

The results show that GPA and HSM have a moderate 0.44 correlation, HSS and HSE have a lower correlation close to 0.30, SATM has an even lower 0.25 correlation and SATV an extremely low 0.11 correlation. In this example, the two SAT scores have a moderately high correlation with each other and the high school grades also correlate well with each other.

Multiple Regression Analysis

1. Open the **csdata** file used in the previous example.

2. Select **Tools** ⇨ **Data Analysis** from the menu and **Regression** from the list of **Analysis Tools.** Click **OK.**

3. Select the GPA data for the **Y Range,** the three high school variables for the **X Range** including the labels, select the **Labels** box, **95% Confidence Level,** the **Output** on a **New Worksheet,** and **Line Fit Plots.** Click **OK.**

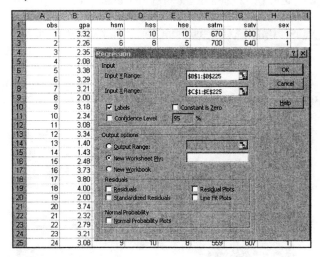

4. The output displays the multiple regression coefficient values under **Coefficients.** The *y*-intercept is 0.59, the HSM coefficient is 0.169, the HSS coefficient is 0.0343, and the HSE coefficient is 0.0451. The resulting equation is

$$GPA = 0.59 + 0.169 * HSM + 0.034 * HSS + 0.045*HSE$$

	A	B	C	D	E	F	G	H	I
1	SUMMARY OUTPUT								
2									
3	*Regression Statistics*								
4	Multiple R	0.452299935							
5	R Square	0.204575231							
6	Adjusted R Square	0.19372853							
7	Standard Error	0.699839129							
8	Observations	224							
9									
10	ANOVA								
11		*df*	*SS*	*MS*	*F*	*Significance F*			
12	Regression	3	27.71233132	9.237444	18.86059	6.35877E-11			
13	Residual	220	107.7504575	0.489775					
14	Total	223	135.4627888						
15									
16		*Coefficients*	*Standard Error*	*t Stat*	*P-value*	*Lower 95%*	*Upper 95%*	*Lower 95.0%*	*Upper 95.0%*
17	Intercept	0.589876617	0.294243238	2.004724	0.046218	0.00998004	1.169773	0.00998004	1.169773194
18	hsm	0.168566642	0.035492139	4.749408	3.68E-06	0.098618493	0.238515	0.098618493	0.238514791
19	hss	0.034315568	0.03755888	0.913647	0.361902	-0.039705729	0.108337	-0.039705729	0.108336865
20	hse	0.045101825	0.038695854	1.165547	0.24506	-0.031160229	0.121364	-0.031160229	0.121363878

Remarks

The ANOVA *F* statistic is 18.86 with a *P*-value of almost zero. We can therefore conclude that at least one of the three regression coefficients for the high school grades is different from zero in the population regression equation.

The *t*-value for the coefficient of HSM is 4.75 which yields a *P*-value of approximately 0.0001 and we conclude that the regression coefficient for this explanatory variable is significantly different from zero. The *P*-values for the other explanatory variables are not statistically significant.

The interpretation of a low *P*-value for a multiple regression coefficient is that x_1 has no value in predicting *y*, given that x_2 and x_3 are available. That is, HSS is not a good predictor of GPA, given that math and English grades are available for prediction. However, this does not mean that science grades cannot predict GPA. In a simple linear regression without HSM and HSE present, the results could be significant. In other words, the results of inference about a single explanatory variable in multiple regression analysis depend on the other explanatory variables in the model.

Analyzing Residual Plots

Residuals can be examined by plotting them against each explanatory variable. We are looking for unusual patterns or curves in the plot.

1. Select **Tools** ⇨ **Data Analysis** from the menu and **Regression** from the list of **Analysis Tools.** Click **OK.**

2. Select the GPA data for the **Y Range**, the three high school variables for the **X Range** including the labels, select the **Labels** box, **95% Confidence Level,** the **Output** on a **New Worksheet, Residuals, Residual Plots** and **Line Fit Plots.** Click **OK.**

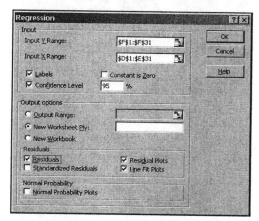

3. The output includes residual plots for HSM, HSS and HSE.

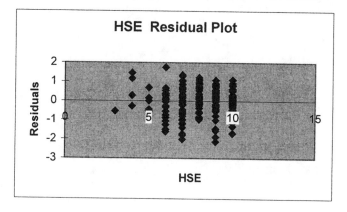

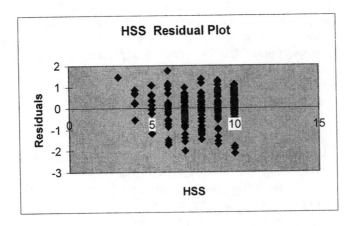

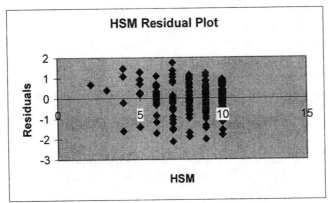

Remarks

Residual plots are used to graphically reveal outliers, influential observations, a curved relationship, clustering, or anything else unusual, which may reveal lurking variables. The residual plots in this exercise are fairly evenly distributed around the center value of zero.

Chapter 12

One-Way Analysis of Variance

In Chapter 7, we compared the means of two data sets in order to infer whether the population means differed significantly. It is also desirable to compare the means of any number of data sets with similar methods to the two-sample t tests.

12.1 Inference for One-Way Analysis of Variance

The statistical methodology for comparing several means is called **analysis of variance**, or **ANOVA**. This analysis is called **one-way** because the populations are compared using a single categorical explanatory variable called a **factor**. For example, it would be appropriate to use this method to compare the tread lifetimes of ten specific brands of tires. But to analyze the effect of more than one variable together, such as tread lifetimes and cost, we would use **two-way ANOVA** methods discussed in Chapter 13.

Comparing Means

The comparison of several means is similar to the comparison of two means. However, instead of a t statistic, ANOVA uses an F-statistic and its P-value to evaluate whether all of the means are equal or differ significantly. The data sets are simple random samples (SRS) from each population or group.

The null hypothesis of ANOVA tests whether the population means are all equal. The alternative is that they are not equal. This could happen if all the means are different or if only one of the means is different. If the null hypothesis is rejected, further analysis needs to be done to determine which population means differ from others.

To determine whether several populations all have the same mean, the variation *among* the means of several groups is compared with the variation *within* groups. Because we are comparing variation, the method is called **analysis of variance**.

The ANOVA Model

Each observation x_j in a random sample of observations from a single normal population with mean μ and standard deviation σ varies about its population mean as

$$x_j = \mu + \varepsilon_j$$

where ε_j are an SRS from the $N(0,\sigma)$ distribution.

Once again, this models **DATA = FIT + RESIDUAL** where **FIT** corresponds to μ and ε corresponds to the **RESIDUAL**. We estimate the population mean μ by the sample mean $\bar{x}$ and the population standard deviation σ by s, the sample standard deviation. The sample residuals are represented by $\varepsilon_j = x_j - \bar{x}$.

The model for one-way ANOVA takes random samples from I different populations. The sample from the ith population has n_j observations, $x_{i1}, x_{i2}, ..., x_{in_i}$ and the one-way model is $x_{ij} = \mu_i + \varepsilon_{ij}$ for $i = 1, ..., I$ and $j = 1, ..., n_j$. The population parameters are the I population means $\mu_1, \mu_2, ..., \mu_j$ and the common standard deviation σ. The standard deviation is assumed to be the same in all of the populations even though sample sizes n_j may differ.

Estimates of Population Parameters

The unknown population parameters in the one-way ANOVA model are the I population means μ_i and the common population standard deviation σ. The population mean μ_i is estimated from the sample mean for the ith group:

$$\bar{x}_i = \frac{1}{n_i} \sum_{j=1}^{n_j} x_{ij}$$

The residuals $\varepsilon_{ij} = x_{ij} - \bar{x}_i$ define the variation about the sample means of the data.

The ANOVA model requires that the population standard deviations are all equal in order for the sample standard deviation to be an estimate of σ. The rule for using this method states that if the largest sample standard deviation is less than twice the smallest sample standard deviation, we can use methods based on the condition that the population standard deviations are equal and the result will be approximately correct.

If we have sample variances $s_1^2, s_2^2, ..., s_I^2$ from I independent SRSs of size $n_1, n_2, ..., n_I$ from populations with common variance σ^2, the pooled sample variance is defined by

$$s_p^2 = \frac{(n_1 - 1)s_1^2 + (n_2 - 1)s_2^2 + ... + (n_I - 1)s_I^2}{(n_1 - 1) + (n_2 - 1) + ... + (n_I - 1)}$$

This is an unbiased estimate of σ^2 and the resulting pooled standard error (estimate of σ) is $s_p = \sqrt{s_p^2}$.

Testing Hypotheses in One-Way ANOVA

The null and alternative hypotheses for one-way ANOVA are

$$H_0 : \mu_1 = \mu_2 = \ldots = \mu_I$$
$$H_a : \text{not all } \mu_i \text{ are equal}$$

Once again, the best way to illustrate this method is by analyzing an example.

IPS Example 12.3 A Study of Workplace Safety

In a study of workplace safety, workers were asked to rate the safety of their work environment and a composite score called the safety climate index (SCI) was calculated. The index is the sum of the responses to ten different questions about safety. The response for each of these questions is an integer ranging from 0 to 10, so the SCI has values from 0 to 100. The workers were classified according to their category as unskilled, skilled, and supervisor.

First, we will need to put the data in an easier format for Excel to analyze.

1. Open the file *eg12_001*.

2. Copy and paste the job categories into separate columns as shown below.

	A	B	C	D	E	F	G
1	jobcat	jobc	SCI		Unskilled	Skilled	Super
2	unskill	1	76		76	78	92
3	unskill	1	61		61	31	81
4	unskill	1	56		56	60	100

3. To investigate whether the standard deviations follow the rule, select **Tools** ⇨ **Data Analysis** ⇨ **Descriptive Statistics.** Select all three columns (Unskilled, Skilled, and Super) including labels, select **Labels in First Row**, select an **Output Range** to the right of the data set, select **Summary Statistics** and **Confidence Level** and **OK.**

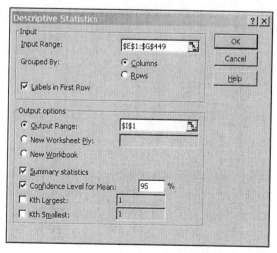

4. Resize the columns. You resulting statistics are shown in three groupings.

I	J	K	L	M	N
Unskilled		*Skilled*		*Super*	
Mean	70.42411	Mean	71.20879	Mean	80.5098
Standard Error	0.863142	Standard Error	1.973915	Standard Error	2.041217
Median	71	Median	72	Median	81
Mode	71	Mode	72	Mode	92
Standard Deviation	18.26926	Standard Deviation	18.82995	Standard Deviation	14.5772
Sample Variance	333.766	Sample Variance	354.567	Sample Variance	212.4949
Kurtosis	-0.06367	Kurtosis	-0.44806	Kurtosis	-0.19515
Skewness	-0.38583	Skewness	-0.41062	Skewness	-0.68292
Range	100	Range	75	Range	54
Minimum	0	Minimum	25	Minimum	46
Maximum	100	Maximum	100	Maximum	100
Sum	31550	Sum	6480	Sum	4106
Count	448	Count	91	Count	51
Confidence Level(95.0%)	1.69632	Confidence Level(95.0%)	3.921524	Confidence Level(95.0%)	4.099907

Remarks

The standard deviations pass the requirement that the largest sample standard deviation be less than twice the smallest sample standard deviation.

The means can be compared by plotted the mean values on a scatterplot.

5. Copy and paste the means to an area below the data as shown below.

Unskilled	70.42411
Skilled	71.20879
Super	80.5098

6. Select both the labels and the mean values and select the **ChartWizard** and then the **Line Graph** (Line Graph is used in this case in order for the labels to be shown on the *x* axis). Click **Finish** and resize the graph.

7. Right click on a gridline and select **Clear.** Click on the series legend and press **Delete.** The resulting scatterplot is shown below.

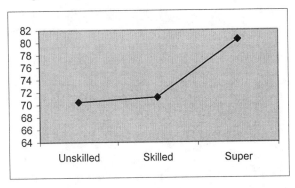

Remarks

It appears that the means for the unskilled workers and the skilled workers are similar, while the supervisors have a higher mean value. We will need to conduct a test in order to determine if this difference is significant.

Histograms can be created for these data sets. The same bins were used to facilitate creating the histograms. The resulting distributions are not shown below and can be considered to be sufficiently normal for ANOVA analysis.

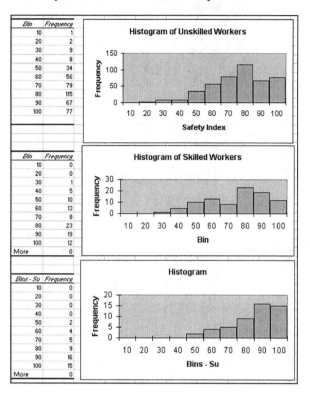

IPS Example 12.6 A Study of Workplace Safety—ANOVA Analysis

We will conduct the ANOVA analysis to determine if the difference in the means is significant.

1. Open the file *eg12_001* used previously. The data should be copied and pasted into separate columns, as seen on p. 189.

2. Select **Tools** ⇨ **Data Analysis** ⇨ **ANOVA: Single Factor** from the menu.

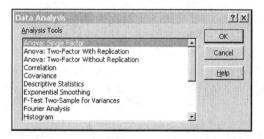

3. Select the entire data set including labels, select **Labels in First Row** and **New Worksheet** for **Output,** and click **OK**.

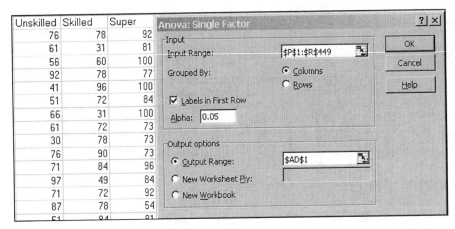

Unskilled	Skilled	Super
76	78	92
61	31	81
56	60	100
92	78	77
41	96	100
51	72	84
66	31	100
61	72	73
30	78	73
76	90	73
71	84	96
97	49	84
71	72	92
87	78	54

3. The resulting output is placed on the same worksheet or a new worksheet. Adjust the width of the columns to see clearly all data labels in the output.

Anova: Single Factor

SUMMARY

Groups	Count	Sum	Average	Variance
Unskilled	448	31550	70.42411	333.766
Skilled	91	6480	71.20879	354.567
Super	51	4106	80.5098	212.4949

ANOVA

Source of Variation	SS	df	MS	F	P-value	F crit
Between Groups	4662.233	2	2331.116	7.136969	0.000866	3.011074
Within Groups	191729.2	587	326.6255			
Total	196391.4	589				

Remarks

The *F*-value is 7.14 with a resulting *P*-value of 0.00086. Therefore, a mean difference would occur by chance 0.09% of the time when the population means are equal. Therefore, there is sufficient evidence that would cause us to reject the null hypothesis that the three populations have equal means.

The ANOVA Table

The Excel software output using both methods is labeled Source of Variation, SS, df, MS, F, *P*-value, and F crit. The row output is labeled Between Groups, Within Groups, and Total. The groups are three sources of variation in the one-way ANOVA (Group identifies the three explanatory variables or "treatments"). The Between Groups row identifies the variation among group means (the FIT term in DATA = FIT + RESIDUAL). The Within Groups row identifies the RESIDUAL of error or variation within a group. The Total corresponds to the resulting DATA term in our model. In other words, the total variation = variation among groups + variation within groups.

Using the Excel built-in analysis tool, the summary is shown below:

ANOVA						
Source of Variation	SS	df	MS	F	P-value	F crit
Between Groups	4662.233	2	2331.116	7.136969	0.000866	3.011074
Within Groups	191729.2	587	326.6255			
Total	196391.4	589				

The variation is expressed by sums of squares (SS). Each sum of squares is equal to the sum of the squares of a set of deviations about their overall mean ($x_{ij} - \bar{x}$), or SST. This equals the variation of the group mean about the overall mean ($\bar{x}_i - \bar{x}$), or SSG, added to the sum of squares of deviations of each observation from its group mean ($x_{ij} - \bar{x}_i$), or SSE. The result is SST = SSG + SSE.

In the above outputs, SST = 196,391.4, SSG = 4662.2, and SSE = 191,729.2. In this example, most of the variation is coming from the error within groups.

Degrees of freedom are defined for each row. SST measures the variation in all N observations about the overall mean, therefore, its degree of freedom or DFT = $N - 1$ or 589. SSG measures the variation of I sample means about the overall mean and has degrees of freedom DFG = $I - 1$ or 2. SSE is the sum of squares of the deviations where N observations are being compared with I sample means resulting in DFE = $N - I$ or 587. The result is DFT = DFG + DFE.

The MS column corresponds to the mean squares for the three sources. The mean square corresponding to the total source is the sample variance calculated assuming that we have one sample from a single population. The calculation is the result of dividing the sum of squares (SS) column by the degrees of freedom (df). The total mean square value is not shown. The calculated F-statistic, the resulting P-value, and the critical F-value are shown.

If H_0 is true, there are no differences among the group means, and the two calculated mean squares should be almost equal. The F-statistic is calculated by dividing the mean squares of the two given sources. In this case, the F-statistic = 2331.1/326.5 = MSG/MSE = 7.14. The P-value of the F test is the probability that a random variable having the $F(I - 1, N - I)$ distribution is greater than or equal to the calculated value of the F statistic.

A summary of the calculations is shown in the following table:

Source	Degrees of Freedom	Sum of Squares	Mean Square	F
Groups	$I - 1$	$\sum_{groups} n_i(\bar{x}_i - \bar{x})^2$	SSG/DFG	MSG/MSE
Error	$N - I$	$\sum_{groups} (n_i - 1)s_i^2$	SSE/DFE	
Total	$N - 1$	$\sum_{obs} (\bar{x}_{ij} - \bar{x})^2$	SST/DFT	

One other calculation is referred to as the coefficient of determination and defined as

$R^2 = \dfrac{SSG}{SST}$. This value compares to R in a multiple regression.

In our safety example, $R^2 = \dfrac{4662.3}{196,391.4} = 0.02$. This means that about two percent of the variation in SCI scores is explained by being part of the unskilled workers, skilled workers, or supervisors. The other 98 percent of the variation is due to variation within each of the three groups, illustrated by the histograms of the distributions of each category.

12.2   Comparing the Means

The **ANOVA F test** tells us if the differences among the observed group means are significant. However, this could be the result of all means differing or only one mean differing from the rest. Plotting and analyzing the data sets can give insight into further analysis of the results. Our hypotheses are

$$H_0: \mu_B = \mu_D = \mu_S$$
$$H_a: \mu_B \neq \mu_D \neq \mu_S$$

If the ANOVA F test rejects the null hypothesis, additional analysis is required to determine whether all means or only one mean is significantly different than the others.

For example, does the data provide any evidence to support a conclusion that the unskilled workers and the skilled workers have different mean SCI scores? The hypotheses associated with this investigation are

$$H_0: \mu_{UN} = \mu_{SK}$$
$$H_a: \mu_{UN} \neq \mu_{SK}$$

These combinations of means are called **contrasts**. Contrast analysis is not easily accomplished in Excel and will not be covered in this manual. Instead, we will use t procedures for testing the differences between two means multiple times to compare all pairings of groups.

Using t Procedures for Multiple Comparisons

IPS Example 12.16 Worker Safety – Multiple Comparisons

The first hypotheses associated with this investigation are

$$H_0: \mu_{UN} = \mu_{SK}$$
$$H_a: \mu_{UN} \neq \mu_{SK}$$

We will conduct a *t* test on the two samples following the procedures learned in Chapter 7. There is one discrepancy with using this method. Each statistic should use the pooled estimator from all groups rather than the pooled estimator from just the two groups being compared. Our results in the following example will be only approximately accurate for this reason but will give us a reasonable idea of what specific groups are significant from each other.

1. Open the file *eg12_001* used previously. The data should be copied and pasted into separate columns, as seen below.

2. Select **Tools** ⇨ **Data Analysis** ⇨ **t-Test: Two-Sample Assuming Equal Variances** from the menu. This method used a pooled standard deviation.

3. Select the **Variable 1 Range** (Unskilled) and the **Variable 2 Range** (Skilled). Keep the **Hypothesized Mean Difference** as "0". Select the **Labels** box if you selected labels and keep the **Alpha** value at 0.05. Select an appropriate **Output Range** (upper left of a blank area to the right of the data set) and click **OK**.

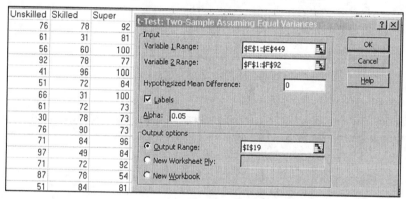

4. Widen the label column of the output. The resulting *P*-value of 1.3% is significant at the 5% level. This method does not calculate the confidence interval.

t-Test: Two-Sample Assuming Equal Variances		
	Unskilled	*Skilled*
Mean	70.42411	71.20879121
Variance	333.766	354.567033
Observations	448	91
Pooled Variance	337.2522	
Hypothesized Mean Difference	0	
df	537	
t Stat	-0.37161	
P(T<=t) one-tail	0.355166	
t Critical one-tail	1.647695	
P(T<=t) two-tail	0.710333	
t Critical two-tail	1.96439	

Remarks and Further Analysis

The *t* test conducted on the means of skilled versus unskilled groups reveal that there is nothing significant in the findings. This is consistent with our previous observation that the means of skilled and unskilled workers were similar.

The next hypotheses associated with this investigation are

$$H_0: \mu_{SK} = \mu_{SU}$$
$$H_a: \mu_{SK} \neq \mu_{SU}$$

5. Conduct another t test to compare the skilled and supervisor groups.

6. The resulting output is shown below.

t-Test: Two-Sample Assuming Equal Variances		
	Skilled	*Super*
Mean	71.20879	80.50980392
Variance	354.567	212.494902
Observations	91	51
Pooled Variance	303.827	
Hypothesized Mean Difference	0	
df	140	
t Stat	-3.05055	
P(T<=t) one-tail	0.001366	
t Critical one-tail	1.65581	
P(T<=t) two-tail	0.002733	
t Critical two-tail	1.977055	

Remarks

The differences in means between the supervisor group and the skilled group are significantly different with a *P*-value of 0.0014. The unskilled group can also be compared to the supervisor group in the same way.

The last hypotheses associated with this investigation are

$$H_0: \mu_{UN} = \mu_{SU}$$
$$H_a: \mu_{UN} \neq \mu_{SU}$$

7. Conduct another t test to compare the skilled and supervisor groups.

8. The resulting output is shown below.

t-Test: Two-Sample Assuming Equal Variances		
	Unskilled	*Super*
Mean	70.42411	80.50980392
Variance	333.766	212.494902
Observations	448	51
Pooled Variance	321.5657	
Hypothesized Mean Difference	0	
df	497	
t Stat	-3.80579	
P(T<=t) one-tail	7.95E-05	
t Critical one-tail	1.647925	
P(T<=t) two-tail	0.000159	
t Critical two-tail	1.96475	

Remarks

The differences in means between the supervisor group and the unskilled group are significantly different with a *P*-value of 0.00008. Our multiple comparison results show that both the skilled and unskilled worker means were significantly different from the supervisor group mean.

Chapter 13

Two-Way Analysis of Variance

In Chapter 7, we compared the means of two data sets in order to infer whether the population means differed significantly. We expanded on this in Chapter 12 to compare the means of several populations according to a single factor in a one-way analysis of variance (ANOVA). This chapter expands this methodology further to compare the means of populations using more than a single factor. We will be introducing the **two-way ANOVA** model.

13.1   The Two-Way ANOVA Model

The statistical methodology for comparing several means is called **analysis of variance**, or **ANOVA**. **One-way** ANOVA compares populations using a single categorical explanatory variable called a **factor**. For example, it would be appropriate to use this method to compare the tread lifetimes of ten specific brands of tires. To analyze the effect of more than one variable together, such as tread lifetimes and cost, we would use **two-way ANOVA** methods discussed in this chapter.

The assumptions for the two-way ANOVA model are that we have independent SRSs from each of $I \times J$ normal populations where I = factor A levels and J = factor B levels. In a two-way design, every level of A is compared with every level of B. The sample size for level i of factor A and level j of factor B is n_{ij}. The total number of observations is

$$N = \sum n_{ij}$$

The population means μ_{ij} may be different but all populations have the same standard deviation σ. The population means and standard deviations are unknown parameters.

If x_{ijk} represents the kth observation from the population having factor A at level i and factor B at level j, the two-way statistical model is

$$x_{ijk} = \mu_{ij} + \varepsilon_{ijk}$$

for $i = 1, ..., I$ and $j = 1, ..., J$ and $k = 1, ..., n_{ij}$. The deviations ε_{ijk} are from the $N(0, \sigma)$ distribution.

The FIT part of the model is the means μ_{ij} and the RESIDUAL part is the deviations from their group means.

The sample mean of the observations in the samples are used to estimate the population mean by using the following relationship:

$$\bar{x}_{ij} = \frac{1}{n_{ij}} \sum_k x_{ijk}$$

The RESIDUAL part of the model contains the unknown σ. The sample variances for each SRS are pooled to estimate σ^2.

$$s_p^2 = \frac{\sum(n_{ij} - 1)s_{ij}^2}{\sum n_{ij} - 1)}$$

As in the one-way ANOVA, the numerator is SSE and the denominator is DFE, which is equal to the total number of observations minus the number of groups. DFE $= N - IJ$. The estimator of σ is s_p.

Main Effects and Interactions

The FIT part of the two-way ANOVA is represented by the population means μ_{ij}. In the two-way problem, we have independent samples from each of $I \times J$ groups and so we can start by thinking of the problem as a one-way ANOVA with IJ groups.

In two-way ANOVA, the terms SSM and DFM consist of terms corresponding to a main effect for A, a main effect for B, and an AB interaction.

$$\text{SSM} = \text{SSA} + \text{SSB} + \text{SSAB}$$
$$\text{DFM} = \text{DFA} + \text{DFB} + \text{DFAB}$$

The term SSA represents the variation among the means for the different levels of factor A. There are I means and DFA $= I - 1$ degrees of freedom. SSB represents variation among the means for the different levels of the factor B with DFB $= J - 1$.

Interactions are more complicated. SSAB is SSM $-$ SSA $-$ SSB, which represents the variation in the model that is not accounted for by the main effects. Its degree of freedom is DFAB $= (IJ - 1) - (I - 1) - (J - 1) = (I - 1)(J - 1)$. These interactions are best studied through examples.

IPS Example 13.7 A Study in Food Portions

There is a general consensus that food portions have been increasing but there is little scientific evidence that documents this change. One study used data from three nationally representative surveys to examine this issue. There were over 63,380 individuals providing data for these studies. Three time points were examined. The table shows the means for the number of calories per portion in soft drinks consumed at home and at restaurants.

Location	1978	1991	1996	Mean
Home	130	133	158	140
Restaurant	125	126	155	135
Mean	127	129	156	137

It is helpful to plot the group means to investigate the differences in the means.

Two-Way ANOVA Pre-Analysis

1. Input the above data into a blank Excel worksheet.

2. Select the cells as a block as shown below, including the Location, years, and values.

	A	B	C	D	E
1	Location	1978	1991	1996	Mean
2	Home	130	133	158	140
3	Restaurant	125	126	155	135
4	Mean	127	129	156	137

3. Click the **ChartWizard** 📖 in the toolbar and select **(XY) Scatter** as **Chart Type** with data points connected by straight lines, and click **Next.**

4. Verify that the **Series** is in **Columns** and the preview chart looks like the one below, and click **Next.**

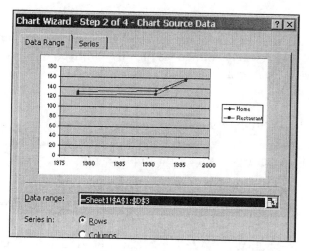

5. Input "Year" for the **Category (X)** axis label and "Calories" for the **Value (Y)** axis label. Click **Finish.**

6. Clear the gridlines by right-clicking on them, and click **Clear.**

7. Changing the scale will enhance the *x*-axis labels. Right-click on the *y* axis, select **Format Axis,** and change the **Minimum** to 110, and click **OK.**

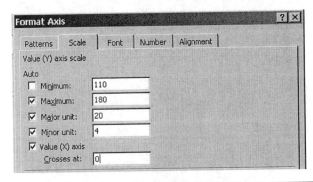

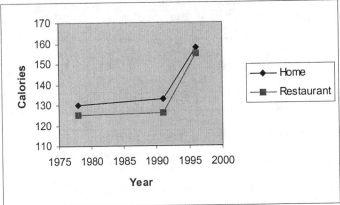

Remarks

In this plot, we can see that the means for 1978 and 1991 are similar but there is a large increase by 1996. This is identified as the main effect of year. There is no clear interaction between location and year.

IPS Example 13.8 A Study in Food Portions - Continued

The survey described in the previous example also obtained data on soft drinks consumed in fast food restaurants. Here is the data from the previous example with the fast food means added:

Location	1978	1991	1996	Mean
Home	130	133	158	140
Restaurant	125	126	155	135
Fast Food	131	143	191	155
Mean	127	129	156	137

1. Add the Fast Food data to the original data from the previous example on your worksheet.

2. Click on the created scatterplot once and click the **ChartWizard** in the toolbar to re-enter the chart selection options. Click **Next** to go to **Step 2** and select the **Series** tab.

3. Click the **Add** button under the Series listing. Type "Fast Food" in the **Name** box. Click inside the **X Values** box and select the years on your worksheet (1978, 1991, 1996). Click inside the **Y Values** box, delete its contents (131, 143, 191), and select the new values for Fast Food. Click **Finish.**

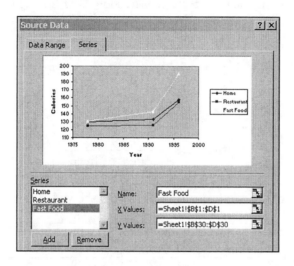

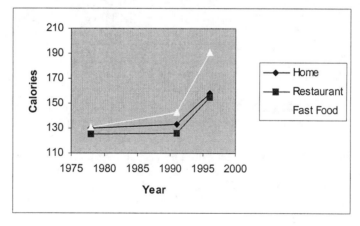

Remarks

Although all three values were close to each other in 1978, Fast Food increased to a higher value than the others in 1990. This trend continues through 1996. This demonstrates an interaction between location and year. However, the main effects are still an important component to the problem. Means are consistently lowest at home, next highest in restaurants, and highest at fast food restaurants.

13.2   Inference for Two-Way ANOVA

The significance of each of the main effects and the interaction is determined by calculating an F statistic for each of the main effects and an additional F statistic for the interaction. The calculations are again organized in an ANOVA table.

The ANOVA Table for Two-Way ANOVA

The results of a two-way ANOVA are summarized in an ANOVA table which splits the total variation SST and the total degrees of freedom DFT among the two main effects and the interaction.

$$SST = SSA + SSB + SSAB + SSE$$
$$DFT = DFA + DFB + DFAB + DFE$$

The mean square is calculated by:

$$\text{mean square} = \frac{\text{sum of squares}}{\text{degrees of freedom}}$$

The general form of the two-way ANOVA table is shown below.

Source	Degrees of Freedom	Sum of Squares	Mean Square	F
A	$I-1$	SSA	SSA/DFA	MSA/MSE
B	$J-1$	SSB	SSB/DFB	MSB/MSE
AB	$(I-1)(J-1)$	SSAB	SSAB/DFAB	MSAB/MSE
Error	$N-IJ$	SSE	SSE/DFE	
Total	$N-1$	SST	SST/DFT	

There are three null hypotheses in the two-way ANOVA, with an F test for each: the main effect of A, the main effect of B, and the AB interaction.

The main effect of factor A is determined by calculating the F statistic:

$$F_A = \frac{MSA}{MSE}$$

where MSA is the mean square of A and MSE is a measure of the variation between the groups.

Similarly, The main effect of factor B is determined by calculating the F statistic:

$$F_B = \frac{MSB}{MSE}$$

where MSB is the mean square of B and MSE is a measure of the variation between the groups.

To test the interaction of A and B, the following F statistic is calculated:

$$F_{AB} = \frac{MSAB}{MSE}$$

IPS Example 13.11 Cardiovascular Risk Factors and Runners

A study of cardiovascular risk factors compared runners who averaged at least 15 miles per week with a control group described as "generally sedentary." Both mean and women were included in the study. The design is a 2×2 ANOVA with the factors group and gender. There were 200 subjects in each of the four combinations. One of the variables measured was the heart rate after 6 minutes of exercise on a treadmill.

1. Copy and paste the data set into four separate columns as shown below, in order to analyze the descriptive statistics.

Control - F	Control -M	Runner -F	Runner - M
159	127	119	100
183	99	84	120
140	157	89	93
140	102	119	107
125	97	127	138
155	122	111	96
148	128	115	107
132	136	109	119
158	142	111	99
136	127	120	102
154	127	105	87
169	157	104	88

2. Select **Tools** ⇨ **Data Analysis** from the menu, select **Descriptive Statistics**, and click **OK**.

3. Select the entire block of data for the **Input Range**, including labels, and select an **Output Range** to the right of the data set. Select **Summary Statistics** and click **OK**.

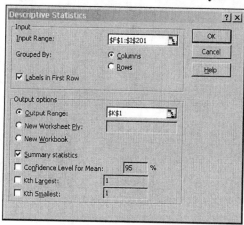

4. Widen the label columns for the output.

Control - F		Control -M		Runner -F		Runner - M	
Mean	148	Mean	130	Mean	115.985	Mean	103.975
Standard Error	1.15053	Standard Error	1.209178	Standard Error	1.129359	Standard Error	0.883843
Median	147	Median	130	Median	116	Median	103
Mode	136	Mode	134	Mode	126	Mode	102
Standard Deviation	16.27095	Standard Deviation	17.10035	Standard Deviation	15.97154	Standard Deviation	12.49942
Sample Variance	264.7437	Sample Variance	292.4221	Sample Variance	255.0902	Sample Variance	156.2356
Kurtosis	0.107601	Kurtosis	0.01151	Kurtosis	-0.14992	Kurtosis	0.21657
Skewness	-0.02531	Skewness	-0.13381	Skewness	0.103511	Skewness	0.314678
Range	91	Range	95	Range	86	Range	77
Minimum	105	Minimum	77	Minimum	78	Minimum	69
Maximum	196	Maximum	172	Maximum	164	Maximum	146
Sum	29600	Sum	26000	Sum	23197	Sum	20795
Count	200	Count	200	Count	200	Count	200

Remarks

There is no indication that there is a violation of the two-way ANOVA condition that all population standard deviations have to be the same. Additional preliminary analysis would include an investigation into normality with histograms and graphing the means.

Two-Way ANOVA Analysis

IPS Example 13.11 Cardiovascular Risk Factors and Runners Continued

A complete analysis of the effects of more than one factor includes a two-way ANOVA output from Excel.

1. Data has to be input into Excel in a very precise format. **Control** and **Runner** are placed in separate columns, and the groupings for **Female** and **Male** are identified in the same column, using labels to the left. Copy and paste the data to the format modeled below.

	Control	Runner
Female	159	119
	183	84
	140	89
	140	119
	125	127
	155	111
	148	115
	132	109
	158	111

2. Select **Tools** ⇨ **Data Analysis** from the menu, select **ANOVA: Two-Factor With Replication**, and click **OK**.

3. Select the **Input Range** as all the data including labels (three columns). Input "200" **Rows per Sample**. Select an **Output Range** to the right of your data set and click **OK**.

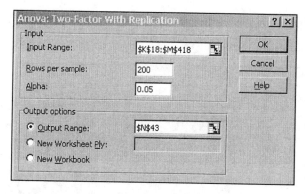

4. The resulting output is shown below:

Anova: Two-Factor With Replication					
SUMMARY	Control	Runner	Total		
Female					
Count	200	200	400		
Sum	29600	23197	52797		
Average	148	115.985	131.9925		
Variance	264.7437	255.0902261	516.1478		
Male					
Count	200	200	400		
Sum	26000	20795	46795		
Average	130	103.975	116.9875		
Variance	292.4221	156.2355528	393.5161		
Total					
Count	400	400			
Sum	55600	43992			
Average	139	109.98			
Variance	359.0877	241.2978446			
ANOVA					
Source of Variation	*SS*	*df*	*MS*	*F*	*P-value*	*F crit*
Sample	45030.01	1	45030.01	185.9799	3.29E-38	3.85316
Columns	168432.1	1	168432.1	695.647	1.1E-110	3.85316
Interaction	1794.005	1	1794.005	7.409481	0.00663	3.85316
Within	192729.8	796	242.1229			
Total	407985.9	799				

Remarks

The top portion of the Excel output provides summary statistics and the lower section is the ANOVA table. There are three possible significance tests due to the two factors (group and gender) and an interaction. The first column lists the four sources of variation. Sample = Factor A, Columns = Factor B, Interaction, and Within = Error. The columns to the right of the first column define the Sums of Squares (SS), degrees of freedom (df), mean square (MS), computed F-values, P-values, and the critical F-values.

All three effects are statistically significant. The group effect is the largest, followed by the gender effect and then the interaction.

The resulting graph of the means illustrates the differences for the main effects clearly. Females have higher heart rates than the men in both groups.

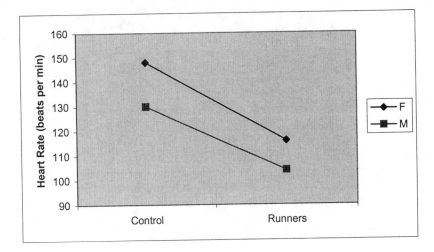

Chapter 14

Bootstrap Methods and Permutation Tests *

The resampling and bootstrap methods described in the text in Chapter 14 use the statistical program S-PLUS. Developing these methods in Excel is time consuming and will not be pursued in this manual.

Chapter 15

Nonparametric Tests

The investigations in previous chapters into various methods of inference have proven to be robust and not very sensitive to a moderate lack of normality, especially when the sample size is fairly large. There are several options for dealing with non-normal distributions. This chapter deals with nonparametric methods, defined as inference procedures that do not require any specific type of population distribution.

15.1   The Wilcoxon Rank Sum Test

One type of nonparametric test is based on the rank (ordered position) of each observation in the data set. The Wilcoxon Rank Sum Test addresses the common two-sample problem.

The rank tests studied in this chapter focus on the center of a population. If a population has a normal distribution, the center is the mean. For a skewed distribution, the center is best represented by the median. The hypotheses for the rank tests will replace the mean with the median as the measure of the center.

Observations are ranked by sorting them in order from smallest to largest, combining observations from both data sets (for a two-sample problem). The rank of each observation is its position in this ordered list, starting with rank 1 for the smallest observation. If an SRS is taken from each population, there are a total of N observations in all, where $N = n_1 + n_2$. The sum W of the ranks for the first sample is the Wilcoxon rank sum statistic. The mean of W is defined by:

$$\mu_W = \frac{n_1(N+1)}{2}$$

The standard deviation is defined by:

$$\sigma_W = \sqrt{\frac{n_1 n_2 (N+1)}{12}}$$

The Wilcoxon rank sum test rejects the hypothesis that the two populations have identical distributions when the rank sum W is far from its mean.

Wilcoxon Rank Sum Test Calculation

IPS Example 15.1 Weeds and Corn Yield

Does the presence of small numbers of weeds reduce the yield of corn? Lamb's-quarter is a common weed in cord fields. A researcher planted corn at the same rate in 8 small plots of ground, then weeded the corn rows by hand to allow no weeds in 4 randomly selected plots and exactly 3 lamb's-quarter plants per meter of row in the other 4 plots. The table below shows the yields of corn (bushels per acre) in each of the plots.

Weeds per meter	Yield (bushels/acre)			
0	166.7	172.2	165.0	176.9
3	158.6	176.4	153.1	156.0

The data set can be investigated for normality by creating stemplots or histograms. In this case, the samples are too small to determine normality adequately or to feel confident in performing the t test. We prefer to use a test that does not require normality.

The hypotheses tested are:

H_0: No difference in the distribution of yields against the one-sided alternative.

H_a: Yields are systematically higher in weed-free plots.

1. Open the file **eg15_001** from the **IPS CD-ROM.**

2. Order the data from smallest to largest by clicking in any cell of the data set and selecting **Data** ⇨ **Sort** from the menu. Click on the down arrow in the **Sort by** box and select **yield.** Verify that the **My List has a Header Row** is selected, and click **OK.**

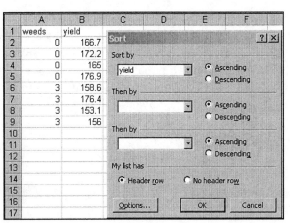

3. Insert a column to the right entitled **Rank.** Number each data point with a ranking from 1 to 8. The data set should look like the one that follows.

	A	B	C
1	weeds	yield	Rank
2	3	153.1	1
3	3	156	2
4	3	158.6	3
5	0	165	4
6	0	166.7	5
7	0	172.2	6
8	3	176.4	7
9	0	176.9	8

4. The sum of rankings can be calculated using the **AutoSum** function and the number of observations by the **Count** function. Create a summary table below the data set, summing the ranks for each treatment category. The resulting values are shown below.

Treatment	n	Sum of Ranks
No Weeds	4	23
Weeds	4	13

5. Calculate the mean of the Wilcoxon rank sum statistic W by inputting the following formula into an empty cell: $=\dfrac{n_1(N+1)}{2}$.

	Treatment	n	Sum of Ranks
11			
12	No Weeds	4	23
13	Weeds	4	13
14			
15	W mean	=C12*(C12+C13+1)/2	

6. Calculate the standard deviation of the Wilcoxon rank sum statistic W by inputting the following formula into an empty cell using cell locations: $=\sqrt{\dfrac{n_1 n_2(N+1)}{12}}$.

	Treatment	n	Sum of Ranks
11			
12	No Weeds	4	23
13	Weeds	4	13
14			
15	W mean	18	
16	W Std Dev	=SQRT(C12*C13*(C12+C13+1)/12)	

7. The resulting value is shown below.

Treatment	n	Sum of Ranks
No Weeds	4	23
Weeds	4	13
W mean	18	
W Std Dev	3.4641	

Remarks

Although the rank sum for males at 235 is higher than the mean, it is only about 1.4 standard deviations higher. Therefore, it does not appear that the data give strong evidence that yields are higher in the population of weed-free corn.

The *P*-value for getting a result ≥ 23 cannot be exactly calculated in Excel but can be approximated by the normal distribution calculation.

8. Calculate the *z* test statistic by inputting the following formula in an empty cell. The first argument in the function is the *x* value, the second argument is the population mean, and the third argument is the standard deviation.

    =STANDARDIZE(23,18,3.464)

9. Press **Enter.** The result is the *z* test statistic.   1.4434

10. Calculate the probability of a given *z* test statistic by inputting the following function in an empty cell.

     =NORMSDIST(1.4434)

11. Press **Enter.** The result is the one-sided probability.   0.9255

12. The given probability represents the area to the left of a given *z*-value. For this example, we want the probability greater than the given value of 1.4434. Input the following formula into an empty cell to calculate the area to the right of 1.4434:

     0.9255
     =1-C19

13. The resulting probability is: 0.0745

14. A more accurate result can be obtained by using a **continuity correction,** testing the entire interval from 22.5 to 23.5 instead of 23, assuming the whole number 23 occupies the entire interval of 22.5 to 23.5. The resulting *P*-Value for a value of 22.5 yields a more accurate result of 0.097, which is close to the exact result from the *W* distribution of $P = 0.10$.

15.2 The Wilcoxon Signed Rank Test

The Wilcoxon Signed Rank Test is a rank test for matched pairs and single samples. If an SRS is taken from each population for a matched pairs study, we will compile a list of the absolute value (without the sign) of the differences in two samples. We will remove all zero differences. The sum W^+ of the ranks for the positive differences is the Wilcoxon signed rank statistic. The mean of W^+ is defined by:

$$\mu_{W^+} = \frac{n(n+1)}{4}$$

The standard deviation is defined by:

$$\sigma_W = \sqrt{\frac{n(n+1)(2n+1)}{24}}$$

The Wilcoxon signed rank test rejects the hypothesis that there are no systematic differences within pairs when the rank sum W^+ is far from its mean. The test statistic is the sum of the ranks of the positive differences.

IPS Example 15.8 Children Retelling a Story

A study of early childhood education asked kindergarten students to retell two fairy tales that had been read to them earlier in the week. Each child told two stories. The first had been read to them, and the second had been read but also illustrated with pictures. An expert listened to a recording of the children and assigned a score for certain uses of language. Here are the data for five "low-progress" readers in a pilot study:

Child	1	2	3	4	5
Story 2	0.77	0.49	0.66	0.28	0.38
Story 1	0.40	0.72	0.00	0.36	0.55
Difference	0.37	−0.23	0.66	−0.08	−0.17

We wonder if illustrations improve how the children retell a story. We would like to test the hypotheses:

H_0: Scores have the same distribution for both stories.
H_a: Scores are systematically higher for Story 2.

Because these are matched pairs data, we base our inference on the differences.

1. Open **eg15_008** from the **IPS CD-ROM**.

2. Calculate the absolute value of the difference of the two data sets and place it in column C. Use the **ABSOLUTE** function. The syntax is shown in the example that follows.

D2		=	=ABS(B2-C2)
	A	B	C	D
1	child	story1	story2	Difference
2	1	0.77	0.4	0.37
3	2	0.49	0.72	0.23
4	3	0.66	0	0.66
5	4	0.28	0.36	0.08
6	5	0.38	0.55	0.17

3. Rank the absolute value of the differences from lowest to highest and assign a rank to each.

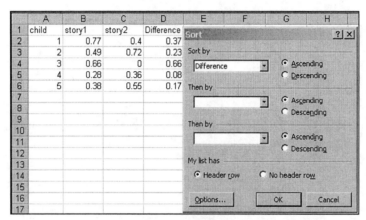

	A	B	C	D	E
1	child	story1	story2	Difference	Rank
2	4	0.28	0.36	0.08	1
3	5	0.38	0.55	0.17	2
4	2	0.49	0.72	0.23	3
5	1	0.77	0.4	0.37	4
6	3	0.66	0	0.66	5

4. Calculate the mean of the Wilcoxon signed rank statistic W^+ by inputting the following formula into an empty cell: $= \dfrac{n(n+1)}{4}$.

	=	=5*(5+1)/4	
B	C	D	E
story1	story2	Difference	Rank
0.28	0.36	0.08	1
0.38	0.55	0.17	2
0.49	0.72	0.23	3
0.77	0.4	0.37	4
0.66	0	0.66	5
		W mean	7.5

5. Calculate the standard deviation of the Wilcoxon signed rank statistic W^+ by inputting the following formula into an empty cell using cell locations where appropriate:
$$= \sqrt{\frac{n(n+1)(2n+1)}{24}}.$$

=	=SQRT(5*(5+1)*(2*5+1)/24)	
C	D	E
story2	Difference	Rank
0.36	0.08	1
0.55	0.17	2
0.72	0.23	3
0.4	0.37	4
0	0.66	5
	W mean	7.5
	W Std Dev	3.708

6. The one-sided *P*-value for $P(W^+ \geq 9)$ is approximated by the normal distribution calculation using the continuity correction as shown in the previous example.

t statistic	=(8.5-7.5)/(E9)
P-value	=TDIST(I19,5,1)

t statistic	0.2697
P-value	0.3991

Remarks

The resulting *P*-Value is close to the actual *P*-value of 0.4062 calculated using statistical software. The result gives no evidence that seeing illustrations improves the storytelling of low-progress readers.

Chapter 16

Logistic Regression

In Chapters 10 and 11, we studied simple and multiple linear regression methods used to model the relationship between a quantitative response variable and one or more explanatory variables. This chapter deals with a similar method for use when the response variable has only two possible values—"success" or "failure" as defined by the problem. In addition, an explanatory variable x will define how a proportion of successes depend on a particular variable. Logistic regression is a statistical method for describing these types of relationships.

16.1 The Logistic Regression Model

Data for logistic regression are n independent observations, each representing an x variable defined as "success" or "failure" for that trial, represented by 1 for a "success" and 0 for a "failure." The mean is identified as the proportion of ones or P(success). This method differs from the Binomial method in that the probability p of a success depends on the value of x. If we are interested in whether a customer makes a purchase or not after being offered a discount, p is the probability that the customer makes a purchase depending on a number of explanatory variables that can be either categorical or quantitative—customer age, type of discount, gender, whether the customer has made a purchase in the past. Logistic regression defines these types of relationships. We will start with an example where there are only two values of p, one for each value of x. In this case, the count of successes has a binomial distribution.

We cannot perform logistic regression analysis in Excel; however, we can understand a few of the basic concepts on the model by using the some of the worksheet tools. Complete analysis should be done in a statistical program such as SPSS, SPS, or Minitab.

IPS Example 16.1 Binge Drinkers

Example 8.1 describes a survey of 17,096 students in U.S. four-year colleges. The researchers were interested in estimating the proportion of students who are frequent binge drinkers. A student who reports drinking five or more drinks in a row three or more times in the past two weeks is called a frequent binge drinker. In the notation of Chapter 5, p is the proportion of frequent binge drinkers in the entire population of college students in U.S. four-year colleges. The number of frequent binge drinkers is an SRS of size n with a binomial distribution with parameters n and p. One promising explanatory variable is the gender of the student. The gender is expressed numerically as 1 if the student is a man and 0 if the student is a woman.

The sample size is $n = 17{,}096$, and the number of frequent binge drinkers in the sample is 3314. The sample contained 7180 men and 9916 women. The probability that a randomly chosen student is a frequent binge drinker has two values, p_1 for men and p_0 for women.

Proportion and Odds Analysis

1. Input the data set into an empty worksheet. Calculate the **proportions** for men and women binge drinkers and total binge drinkers as shown below:

	D2	▼	=	=B2/C2
	A	B	C	D
1		Binge D	Total	Prop
2	Men	1630	7180	0.2270
3	Women	1684	9916	0.1698
4	Total	3314	17096	0.1938

2. Logistic regression works with odds rather than proportions. The odds are the ratio of the proportions for the two possible outcomes.

$$\text{ODDS} = \frac{p}{1-p} = \frac{\text{probabiliy of success}}{\text{probability of failure}}$$

Calculate the estimated odds of each group and the total of being a frequent binge drinker.

	E4	▼	=	=D4/(1-D4)	
	A	B	C	D	E
1		Binge D	Total	Prop	Odds
2	Men	1630	7180	0.2270	0.2937
3	Women	1684	9916	0.1698	0.2046
4	Total	3314	17096	0.1938	0.2405

Remarks

Odds are often rounded to integers. The odds that a woman is a frequent binge drinker is 0.205, which is approximately 1 in 5. The odds that a woman is NOT a frequent binge drinker are 5 to 1.

Model for Logistic Regression

Simple linear regression models the mean μ of the response variable y as a linear function of the explanatory variable x: $\mu = \beta_0 + \beta_1 x$. Logistic regression models the mean of the response variable p in terms of an explanatory variable x. The logistic regression solution will transform the odds using a **natural** logarithm. This will ensure that $0 \le p \le 1$.

$$\log\left(\frac{p}{1-p}\right) = \beta_0 + \beta_1 x$$

IPS Examples 16.3 & 16.4 Binge Drinkers, continued

For our binge-drinking example, there are $n = 17{,}096$ students in the sample. The explanatory variable is gender, which has been coded with an indicator variable with values $x = 1$ for men and $x = 0$ for women. The response variable is also an indicator variable. Thus, the student is either a frequent binge drinker or the student is not a frequent binge drinker. The model says that the probability (p) that this student is a frequent binge drinker depends upon the student's gender ($x = 1$ or $x = 0$). There are two possible values for p, p_{men} and p_{women}.

1. Add a column to the previous table to calculate log(ODDS). The function LN is used to calculate the natural logarithm of the odds fraction for both genders and the total.

F2	▾	=	=LN(D2/(1-D2))		
	A	B	C	D	E	F
1		**Binge D**	**Total**	**Prop**	**Odds**	**Log Odds**
2	**Men**	1630	7180	0.2270	0.2937	-1.225218
3	**Women**	1684	9916	0.1698	0.2046	-1.586857
4	**Total**	3314	17096	0.1938	0.2405	-1.425207

The log(OSSA) for men is $y = \log\left(\dfrac{\hat{p}_{men}}{1-\hat{p}_{men}}\right) = -1.23$. The log(ODDS) for women is

$y = \log\left(\dfrac{\hat{p}_{women}}{1-\hat{p}_{women}}\right) = -1.59$.

The estimate of the intercept b_0 is the log(ODDS) for women $= -1.59$. The slope is the difference between the log(ODDS) for the men and the log(ODDS) for the women:

$$b_1 = -1.23 - (-1.59) = 0.36.$$

The fitted logistic regression model is

$$\log(\text{ODDS}) = -1.59 + 0.36x$$

The **odds ratio**, that is, the ratio of the odds that a man is a frequent binge drinker to the odds that a man is a frequent binge drinker is shown as

$$\frac{\text{ODDS}_{men}}{\text{ODDS}_{women}} = e^{0.36} = 1.43$$

These formulas can be input into our Excel table as shown below:

	A	B	C	D	E	F	G	H	I
1		Binge D	Total	Prop	Odds	Log(ODDS)	Intercept	b1	ODDS Ratio
2	Men	1630	7180	=B2/C2	=D2/(1-D2)	=LN(D2/(1-D2))	=F3	=F2-(F3)	=EXP(0.36)
3	Women	1684	9916	=B3/C3	=D3/(1-D3)	=LN(D3/(1-D3))			
4	Total	=SUM(B2:B3)	=SUM(C2:C3)	=B4/C4	=D4/(1-D4)	=LN(D4/(1-D4))			

The results are shown below:

	A	B	C	D	E	F	G	H	I
1		Binge D	Total	Prop	Odds	Log(ODDS)	Intercept	b1	ODDS Ratio
2	Men	1630	7180	0.2270	0.2937	-1.23	-1.59	0.36	1.43
3	Women	1684	9916	0.1698	0.2046	-1.59			
4	Total	3314	17096	0.1938	0.2405	-1.43			

Remarks

The resulting odds for mean are 1.43 times the odds for women. If women had been coded as 1 and men as 0, the signs of the parameters would be reversed and the equation would be log(ODDS) = 1.59 − 0.36x. The odds ratio would be $e^{-0.36} = 0.70$. The odds for women are 70% of the odds for men.

16.2 Inference for Logistic Regression

Statistical inference for logistic regression is similar to statistical inference for simple linear regression. The equations are complicated to input into Excel. This type of analysis is best conducted using a statistical program.

Chapter 17

Statistics for Quality: Control and Capability

Statistical process control is important in manufacturing as a method to keep the observed pattern of values stable or constant over time. There is always variation but putting statistical limits on those variations keeps the process (a chain of activities that turns inputs into outputs) under control.

17.1 Processes and Statistical Process Control

The goal of statistical process control is to make a process stable over time and then keep it stable. All processes have variations, however, if the pattern of variation remains stable, the process control will also remain stable. Each of these processes is made up of several successive activities that eventually produce an output.

$\bar{x}$ Charts for Process Monitoring

A quantitative variable x that has a normal distribution is being measured in a process that has been operating in control for a long period. The process mean μ and the process standard deviation σ describe the distribution of x as long as the process remains in control. In actuality, the process mean and standard deviation have to be estimated from past data. The law of large numbers lends support to the validity of using the past data. The mean μ and the standard deviation σ describe the center and the spread of our variable x, as long as the process remains in control.

The means of successive samples are plotted on the $\bar{x}$ chart. Samples of size n are taken during the operation of the process at regular intervals. The means of these samples are plotted against the order in which the samples were taken. A centerline is drawn with mean μ. Dashed control lines are drawn on the chart at the heights of the 99.7% mark or $\mu \pm 3\sigma/\sqrt{n}$. These define the control limits for the process.

> The upper control limit is $\text{UCL} = \mu + 3\sigma/\sqrt{n}$
>
> The lower control limit is $\text{UCL} = \mu - 3\sigma/\sqrt{n}$

where n is the number of samples taken at a particular time interval.

If the process remains in control, the process mean and standard deviation would not change and there would rarely be an observation $\bar{x}$ that would lie outside the control limits. If such an observation were observed, it would be an indication that the process had been disturbed or altered in some way.

IPS Example 17.4 Manufacturing Computer Monitors

A manufacturer of computer monitors must control the tension on the mesh of fine vertical wires that lies behind the surface of the viewing screen. Too much tension will tear the mesh and too little will allow the wrinkles. Tension is measured by an electrical device with output readings in millivolts (mV). The manufacturing process has been stable with mean tension $\mu = 275$ mV and process standard deviation $\sigma = 43$ mV.

The mean 275 mV and the common cause variation measured by the standard deviation 43 mV describe the stable state of the process. If these values are not satisfactory, the manufacturer must make some fundamental changes in the process. One advantage of statistical control is that we know how the process should behave and can decide whether its behavior is satisfactory or not. In fact, the common cause variation in mesh tension does not affect the performance of the monitors. We need to watch the process and maintain its current condition.

The operator measures the tension on a sample of 4 monitors each hour. Table 17.1 data on the **IPS CD-ROM** contains the last 20 samples, representing 20 hours of data. The table also contains the mean $\bar{x}$ and the standard deviation s for each sample.

Creating $\bar{x}$ Control Charts

1. Open the file **ta17_001** from the **IPS CD-ROM.**

2. Copy and paste the sample numbers and the means to new columns to the right of the data set. Calculate the control limits as shown and input the center line as 275. Copy all formulas and the 275 down to sample 20.

	I	J	K	L	M
1	Sample	Mean	LCL	UCL	Center Line
2	1	253.4	=M2-3*43/SQRT(4)	=M2+3*43/SQRT(4)	275
3	2	285.4	=M3-3*43/SQRT(4)	=M3+3*43/SQRT(4)	275
4	3	255.3	=M4-3*43/SQRT(4)	=M4+3*43/SQRT(4)	275
5	4	260.8	=M5-3*43/SQRT(4)	=M5+3*43/SQRT(4)	275
6	5	272.7	=M6-3*43/SQRT(4)	=M6+3*43/SQRT(4)	275
7	6	245.2	=M7-3*43/SQRT(4)	=M7+3*43/SQRT(4)	275
8	7	265.2	=M8-3*43/SQRT(4)	=M8+3*43/SQRT(4)	275
9	8	265.6	=M9-3*43/SQRT(4)	=M9+3*43/SQRT(4)	275
10	9	278.5	=M10-3*43/SQRT(4)	=M10+3*43/SQRT(4)	275
11	10	285.4	=M11-3*43/SQRT(4)	=M11+3*43/SQRT(4)	275

3. Select the entire new data set and click on the **ChartWizard** 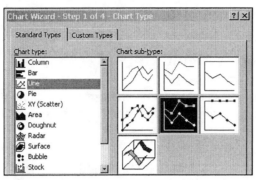 in the menu. Select
    XY (Scatter) graph with points connected with lines. Click Next. **PBS Stats** ⇨ from
    the menu.

4. The preview should look approximately like the one below. Click **Next.**

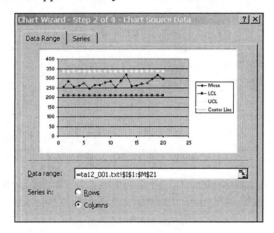

5. Input "Sample Number" for the *x*-axis label and "Sample Mean" for the *y*-axis label.
    Click **Finish.**

6. Clear the gridlines and adjust the *y*-axis scale to focus on the chart. The data series
    for the lines can be reformatted to look more like lines, and not connected points.

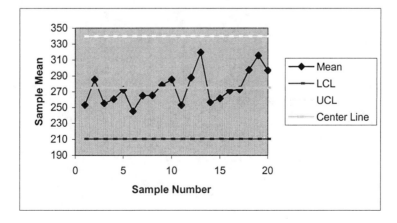

Remarks

The means of the 20 samples vary within the control limits specified indicating a stable process.

s Charts for Process Monitoring

The *s* chart, or sample standard deviation chart, is not as straightforward in design as the $\bar{x}$ chart. The sampling distribution *s* for samples from a normally distributed process is defined by:

The mean of *s* (μ_s) is a constant c_4 times the process standard deviation σ.

The standard deviation (σ_s) of *s* is a constant c_5 times the process standard deviation σ.

The values of these constants depend on the size of the samples. For large samples, the sample standard deviation is an accurate predictor of the process standard deviation σ and, therefore, c_4 is approximately 1.

The three-sigma *s* control chart is constructed by placing the centerline at μ_s and the control limits at $\mu_s \pm 3\sigma_s = c_4\sigma \pm 3c_5 = (c_4 \pm 3c_5)\sigma$. The upper limit constant $= B_6$ and the lower limit constant $= B_5$. All of the control chart constants depend on the sample size n.

These charts and the remainder of the procedures discussed in this chapter are not easily created in Excel. It is more efficient to use a statistical program that automatically creates the charts for analysis.